STUDIES IN GAME THEORY
AND MATHEMATICAL ECONOMICS

General Editor: Andrew Schotter (New York University)

Volumes in this Series:
DIFFERENTIAL GAMES AND OTHER GAME-THEORETIC
TOPICS IN SOVIET LITERATURE
by Alfred Zauberman
BIDDING AND AUCTIONING FOR PROCUREMENT
AND ALLOCATION
edited by Yakov Amihud
GAME THEORY AND POLITICAL SCIENCE
edited by Peter C. Ordeshook
ECONOMICS AND THE COMPETITIVE PROCESS
by James H. Case
AUCTIONS, BIDDING, AND CONTRACTING:
USES AND THEORY
*edited by Richard Engelbrecht-Wiggans, Martin Shubik, and
Robert M. Stark*
GAME THEORY FOR THE SOCIAL SCIENCES,
SECOND AND REVISED EDITION
by Hervé Moulin
EIGHTY-NINE EXERCISES WITH SOLUTIONS FROM
GAME THEORY FOR THE SOCIAL SCIENCES,
SECOND AND REVISED EDITION
by Hervé Moulin

ABOUT THIS SERIES

Game theory, since its creation in 1944 by John von Neumann and Oskar Morgenstern, has been applied to a wide variety of social phenomena by scholars in economics, political science, sociology, philosophy, and even biology. This series attempts to bring to the academic community a set of books dedicated to the belief that game theory can be a major tool in applied science. It publishes original monographs, textbooks and conference volumes which present work that is both of high technical quality and pertinent to the world we live in today.

GAME THEORY FOR THE SOCIAL SCIENCES, SECOND AND REVISED EDITION

Hervé Moulin

NEW YORK UNIVERSITY PRESS
WASHINGTON SQUARE, NEW YORK
1986

Copyright © 1986 by Hermann, 293 rue Lecourbe, 75015 Paris
Manufactured in the United States of America

Library of Congress Cataloging-in-Publication Data

Moulin, Hervé.
Game theory for the social sciences.

(Studies in game theory and mathematical
economics)
Translation of: Théorie des jeux pour l'économie
et la politique.
Bibliography: p.
Includes index.
1. Social sciences—Mathematical models.
2. Game theory. I. Title. II. Series.
H61.25.M6813   1986   300'.1'5193   86-5439
ISBN 0-8147-5430-9
ISBN 0-8147-5431-7 (pbk.)

The medallion on the cover of this series was designed
by the French contemporary artist Georges Mathieu as one of
a set of medals struck by the Musée de la Monnaie in 1971.
Eighteen medals were created by Mathieu to "commemorate 18
stages in the development of western consciousness." The
Edict of Milan in 313 was the first; Game Theory, 1944, was
number seventeen.

FOREWORD
TO THE SECOND AND REVISED EDITION

Game theory is among the most important mathematical tools used in the social sciences today. This book intends to meet two main goals:

i) to offer the non-specialist a self-contained exposition of the essential concepts of strategic games. The definition and results are entirely rigorous from a mathematical standpoint; however more space is devoted to describe and compare the various equilibrium notions than to explore methods for computing them. Technicalities are mostly discarded. Ninety-one exercises, some of them difficult, are offered to the motivated reader.

ii) to justify the usefulness of game theoretical concepts in economic and political theory. For that purpose 43 examples have been carefully chosen for their illustrative power. Their inspiration is mainly microeconomic, with applications to imperfect competition, public goods, voting models, division methods, and so on. In many of our problems, the reader is asked to develop further interpretative arguments from the game model. For these some familiarity with economic reasoning will be helpful.

The book deliberately ignores several important subfields of game theory. Differential games are not treated for, so far, they have had little impact on social sciences. Also they require a solid background in ordinary and partial differential equations. Also omitted are the value approaches to games in characteristic form: in my view, they pertain to the non-strategic models of arbitration, and social choice theory is their natural environment. Finally, I do not say a word about games in incomplete information. This is because, despite the growing applications of these models in economics and politics, the results still lack the maturity and simplicity required in a textbook presentation.

ACKNOWLEDGMENTS

This book took shape while I taught game theory at the Université Paris 9, the Ecole Nationale de ia Statistique et Administration Economique (Paris), and Princeton University, among other places. My first debt is to the students therein. In the preparation of this course, encouragement by Paul Champsaur and Jean Pierre Aubin were most precious. Editorial advice and critical reviews by Andrew Schotter greatly improved it. Material support was provided by the Laboratoire d'Econométrie de l'Ecole Polytechnique (Paris), the Institute for Mathematical Studies in Social Sciences (Stanford), and the Virginia Polytechnic Institute and State University. Last but not least, the handsome typing of Kerry Dylewski gave to this second edition its final form.

CONTENTS

INTRODUCTION

1. NORMAL-FORM GAMES

A game is a mathematical idealization of collective action: Several individual agents (the players, numbered $i = 1, 2, \ldots, n$) influence a situation (called the outcome of the game) and have differing interests (their utility for the various situations differs).

A normal-form game (also called "game in strategic form") among players $\{1, 2, \ldots, n\}$ endows each player i with a strategy set X_i. Each player i freely picks a strategy $x_i \in X_i$, resulting in an outcome $(x_1, \ldots, x_n)$ from which agent i derives utility $u_i(x_1, \ldots, x_n)$. Hence, the object

$$(X_1, \ldots, X_n; u_1, \ldots, u_n)$$

where each utility function u_i (also called payoff) maps
X_1 x ... x X_n into R (the set of real numbers).

A crucial feature of this model is that decisions are
made individually and privately by the players. Once each
player has independently selected a strategy, all relevant
parameters determining individual utility levels are
known. No chance move, no state of nature, no exogenous
action (by an external agent) can influence individual
welfare (measured by utilities) any further. A referee
might be necessary to enforce the rules of the game. He
might be needed to check that each player's strategic
choice is legal (i.e., belongs to X_i) and exact whatever
payments are dictated by the outcome $(x_1, ..., x_n)$. But
he may not influence the actual play of the game by inducing
strategic choices (like the dance-caller or the mediator
in bargaining) or by deciding outcomes based on his own
value judgements (such as, "This outcome seems more fair
to me than that one"). In a game of strategy we think first
of the amoral interaction of several selfish "librearbitres."
Every choice within your strategy set is valid, and your
only value is your own utility. Then, "que le meilleur
gagne."

How will our players exercise their respective strategic
freedom? Can we predict the deterministic outcome of a
game among rational players? Can we at least eliminate
most outcomes as irrational? These are the issues that game

theory is about. They are made difficult because the selfish
interests of our players are typically neither clearly
antagonistic (which would generate unambiguously conflictual
behavior) nor identical (in which case the game would be a
mere coordination problem and best solved by full cooperation
between the actors). In most games inspired by economics or
political science, the configuration of utilities is neither
strict antagonism nor identity. The seller and the buyer
agree that they would both profit from an exchange if the
price lies within their respective reservation levels; but
they eagerly compete for the choice of a particular price
within these limits. Similarly, two moderate voters typically
agree to defeat the extremist in an election but struggle
fiercely to support one of two prominent middle-of-the-road
candidates. A moment's reflection will convince the reader
that most social-gamelike situations generate tendencies both
to conflictual and cooperative behavior.

Games, in their normal form, comprise the simplest and
most powerful model to analyze situations involving mixed
motives. They generate a conceptual panoply formalizing
many behavioral scenarios, some of them noncooperative
(Part II) some clearly cooperative (Part III), and some in
between (Chapter 8). In a given normal-form game, all kinds
of equilibria are established, but some more accurately
indicate plausible behavior within the underlying microeconomic model than others.

2. GAME THEORY IN THE SOCIAL SCIENCES

In most of the social sciences, models abound where
strategic considerations are relevant. Economics is the
privileged field of applied game theory. This was originally
suggested by von Neumann and Morgenstern. Yet, the game
approach gained its well-established reputation only after
the Debreu-Scarf theorem was proposed to settle the
cooperative foundations of the competitive equilibrium.
Since then, entire subfields of economic theory (like
imperfect competition, or the economics of incentives) were
developed as a branch of game theory itself (see the survey
paper of Schotter and Schwödiauer [1980]).

Since Farqharson's book ([1969]), strategically oriented
analysis of voting processes have renewed several traditional
themes of political theory. Although the game models used
there are still controversial, there is enough evidence that
potential contributions of the game viewpoint are vast. See
the survey book of Brams [1975] and the recent literature
on implementation (Maskin [1983], Moulin [1983], Peleg [1984]).

Game-oriented thinking is now common in sociology. (See
Crozier and Friedberg [1977].) Clearly the search for
equilibrium concepts idealizing the spectrum from noncooperative
to cooperative behavior bears on fundamental issues of
sociology. For instance, our view of cooperation as being
comprised of nonbinding agreements (Part III) is implicit in

Rousseau's paradigm of the evolution from the "liberté naturelle" to the "liberté civile" (see Rousseau [1755]). See also Durkheim's idealist (as opposed to contractualist) approach to cooperation (Durkheim [1893]). However, the formal game-theoretical models that have been published in current sociological literature are rare or technically very simple. Still, the impact of game theory seems to be irreversible, if only for pedagogical convenience.

Did von Neumann and Morgenstern, when they were formulating game theory, "settle for the first time a scientifically based explanation of the way through from individual behavior to collective behavior?" (Cazelle [1969]). Most contenders of the theory argue that, although the proposed equilibrium concepts sound plausible and give rise to nice interpretations in the examples, no empirical evidence is provided that the players involved in the game you are describing actually follow the rationale behind equilibrium concepts. Besides, any list given here might very well be incomplete.

We reply that axiomatic methods and simple models are needed in social sciences to isolate fundamental concepts: "the axiomatic language in social sciences signals a rising science, not an advanced one; this is the only way to abstract from actual experiences" (Granger [1955]). We view game theory as a means, in economic and political theory, to develop models. Its increasing use in these fields is its unique and sufficient justification. As Hayek points

6

out, "social sciences progress by means of concepts."
Game theory indeed provides flexible concepts each one of
them throwing light on some aspects of social interaction.

3. AN OVERVIEW OF THE BOOK

Part I describes the (very small) class of games where
each player is endowed with an optimal strategy, so that
the outcome of the game among rational players is deterministic. Those are the two-person, zero-sum games with a
value (Chapter 1) and, in general, the inessential games
(Chapter 2). All other games generate conflictual exchanges
of information, such as the competition for the first or
second move, and deterring threats. Part II analyzes
noncooperative equilibrium concepts. We assume first that
no communication is feasible among individuals and describe
several logical arguments by which a player can decide which
strategy he "should" use: dominant strategies, when our
player ignores of the other's utility (Chapter 3), sophisticated equilibrium strategies, when the utilities of all
players are public knowledge (Chapter 4). These strategies
can all be computed by each player independently of the
other's behavior, in a purely atemporal world. However
appealing these concepts are, their mathematical properties
are bad (dominant and sophisticated equilibrium strategies
typically fail to exist). There comes the "second generation"
of noncooperative equilibrium concept, namely Nash's,

relying on some form of communication among players, if only
through their mutually observed behavior (Chapter 5).
Several scenarios justifying Nash's stability property are
proposed, one of them (the Cournot tatonnement) being an
explicit dynamic scenario (Chapter 6). Contrary to the
first-generation concepts, the Nash-equilibrium outcomes
share good mathematical properties. If the players use
randomized independent strategies, existence of such outcomes
is generally guaranteed (Chapter 7).

The motivation of all cooperative scenarios studied
in Part III is that individual strategic freedom may hurt
the common interest: a noncooperative equilibrium outcome
typically fails to achieve Pareto optimality. This dilemma
between establishing stability and efficiency (of which
Part II offers many examples) suggest taht the players would
be better off by cooperating. It does not say what the
cooperative process should be.

To achieve outcomes with Pareto superiority to the
noncooperative equilibrium, we assume that our players enter
nonbinding agreements. That is to say, they can adopt
behavioral scenarios that do not deprive them of the sovereign
right to play any available strategy. These agreements are
then self-enforced by a stability property of the Nash type:
"I am not forced to follow our agreement, but as long as you
guys are faithful to it, I have no incentive to betray."
In fact, the Nash equilibrium itself is the first fundamental

8

example of such a cooperative agreement, and the sequence of concepts introduced in Part III (with the exception of Chapter 9) develop increasingly subtle variants of it.

The feasibility of a nonbinding agreement relies on some enforced limitation on the communication channels. One possibility is that all kinds of communication are banned after the agreement is reached so that individuals cannot be threatened anymore: this is the context in which the correlated equilibria (Chapter 8) emerge. Alternatively, one may require that all deviations from the agreed upon strategies must occur publicly so that mutual threats can be carried out. This leads to a cooperative view of deterrence, to which Chapters 9 and 10, are devoted.

In a binding agreement some players together agree to play towards a particular outcome, and a supreme authority, recognized by all, enforces the respect of this promise. When they sign the agreement, all players actually lose the control of their own strategy; thus the issue of stability of this agreement vanishes because no betrayal, either individual or coalitional, is any longer possible. Since the cooperative process is petrified after the agreement is signed the difficult questions all arise before it is signed. In other words, since conceivable agreements are not discriminated according to their stability, they are distinguished as more or less "just." Here the view of cooperation switches from positive (Which agreements are stable with respect to some information structure?) to normative (Which agreements treat the players fairly, given the existing power structure?). Such issues of justice are not touched upon. For instance, the Shapley value of a game in characteristic form is not presented (although Chapter 9 addresses the stability issue in games in their characteristic form).

4. SOME EPISTEMOLOGICAL COMMONPLACES

The epistemological features common to all social sciences should be kept in the mind by any newcomer to game theory. The most important one is the "universal irruption of value judgments" (Rickert), illustrated here by such notions as Pareto optimality (implicitly taken to be desirable), the search for stable agreements (stability being viewed as a condition necessary to social consensus), and the respect for individual strategic freedom (that should be preserved as much as possible, see Part III).

As a corollary, no statement, whatever its level of generality, is intended to achieve naked objective validity; it tries, rather, to gain subjective approval by the collectivity (i.e., its however small, intended audience). Henceforth Pygmalion's complex, the traditional sin committed by some involved with the physical sciences (i.e., the tendency of some scientists to reshape reality according to their own models, not the other way around), loses its devilish connotation for those in the social sciences. Pygmalion is, after all, just an unconvincing theoretician: "With regard to human affairs, things are as their actors think they are" (Hayek, quoted by Freund [1967]).

To put it differently, the social scientist can no longer invoke a role of neutral observer: "We explain natural phenomena, we understand the psychic ones" (Dilthey, quoted by Freund [1967]). The unavoidable "empathie" (Boudon) of the observer and the observed materializes in a "verständnis" phase of the analysis where the motivations of the actors are internalized by the scientist. Next the output is an "ideal type" (Weber), namely, a specific model describing convincingly a limited phenomenon with no reference to an underlying general, exhaustive theory. We wish to consider each among the dozen equilibrium concepts presented here as a distinct ideal type describing a particular strategic and/or informational aspect of collective action.

If only one statement would remain from this course in strategic game theory, it might be that a given game generates several, not a large number of, equilibrium concepts depending upon the informational and cooperative patterns. Thus a normal form game, our basic mathematical model, proves to be a rich source of strategic scenarios clarifying the logical connections of such primitive notions as cooperative versus selfish behavior, mutual strategic anticipations, tactical and/or cooperative uncertainty, communication by threats and/or promises. Ultimate criticism of the proposed concepts should bear on their relevance to the illustrative examples.

PART I. THE BREAKDOWN OF STRATEGIC DETERMINISM

CHAPTER 1. TWO-PERSON, ZERO-SUM GAMES

1. GAMES WITH AND WITHOUT A VALUE

A two-person, zero-sum game takes the form $(X_1, X_2, u_1, -u_1)$ where X_i is Player i's strategy set (i = 1, 2) and u_1 is the real valued payoff function defined on $X_1 \times X_2$, which Player 1 tries to maximize and which Player 2 tries to minimize. For convenience we write $u_1 = u$ and denote the game as $G = (X_1, X_2, u)$.

The secure utility level to Player 1 is defined as $\alpha_1 = \sup_{x_1 \epsilon X_1} \inf_{x_2 \epsilon X_2} u(x_1, x_2)$ and a strategy $x^*_1 \epsilon X_1$ is called prudent if

$$\inf_{x_2 \epsilon X_2} u(x^*_1, x_2) = \alpha_1$$

Similarly the <u>secure</u> disutility level to Player 2 is

$\alpha_2 = \inf\limits_{x_2 \epsilon X_2} \sup\limits_{x_1 \epsilon X_1} u(x_1, x_2)$ and a strategy $x^*_2 \epsilon X_2$ is prudent if

$$\sup\limits_{x_1 \epsilon X_1} u(x_1, x^*_2) = \alpha_2$$

<u>Definition 1.</u>

For all two-person, zero-sum games G, we have $\alpha_1 \leq \alpha_2$. If $\alpha_1 = \alpha_2$, this number is called the <u>value</u> of G. If $\alpha_1 < \alpha_2$, we say that G has no value.

When a game has a value, prudent strategies of either player are optimal in the following sense.

<u>Theorem 1.</u>

Say that outcome $(x^*_1, x^*_2) \epsilon X_1 \times X_2$ is a <u>saddle point</u> of u if

for all $x_1 \epsilon X_1$ $\quad u(x_1, x^*_2) \leq u(x^*_1, x^*_2) \leq u(x^*_1, x_2)$ $\quad$ (1)

for all $x_2 \epsilon X_2$

If game G has a value v an outcome (x^*_1, x^*_2) is a saddle point of G iff x^*_1 and x^*_2 are prudent. In that case $u(x^*_1, x^*_2) = v$. If G has no value, it has no saddle point either.

<u>Proof.</u>

Suppose $v = \alpha_1 = \alpha_2$ and x^*_i is a prudent strategy of Player i, i = 1, 2. Then, by definition of prudent strategies

$$\sup_{x_1 \epsilon X_1} u(x_1, x^*_2) = v = \inf_{x_2 \epsilon X_2} u(x^*_1, x_2) \tag{2}$$

In particular, $u(x^*_1, x^*_2) \leq v \leq u(x^*_1, x^*_2)$; hence, $v = u(x^*_1, x^*_2)$. Now (2) is just the same as (1).

Conversely suppose (x^*_1, x^*_2) is a saddle point of G. By (2), this is equivalent to

$$\sup_{x_1 \epsilon X_1} u(x_1, x^*_2) = u(x^*_1, x^*_2) = \inf_{x_2 \epsilon X_2} u(x^*_1, x_2) \tag{3}$$

By definition of α_1, we have $\sup_{x_1 \epsilon X_1} u(x_1, x^*_2) \geq \alpha_2$ and similarly $\inf_{x_2 \epsilon X_2} u(x^*_1, x_2) \leq \alpha_1$. Thus (3) implies $\alpha_2 \leq u(x^*_1, x^*_2) \leq \alpha_1$. Since $\alpha_1 \leq \alpha_2$ always holds, this means G has a value v and $u(x^*_1, x^*_2) = v$. In view of (3), x^*_i is then a prudent strategy of Player i, i = 1, 2. QED.

In a (two-person, zero-sum) game with a value, prudent strategies deserve to be called optimal because of the saddlepoint property (1): Each player guarantees himself of at least the utility v (or at most the disutility v) by using a prudent strategy. To find out his own prudent strategies, a player needs only to solve a decentralized program, e.g., for Player 1:

$$\max_{x_1 \epsilon X_1} \phi(x_1)$$

where $\phi(x_1) = \inf_{x_2 \epsilon X_2} u(x_1, x_2)$. Of course, this program needs not to have a solution, even though G has a value, for obvious topological reasons. (The real-valued function ϕ

does not need to reach its maximum over an infinite set X_1. See however Remark 1 below.)

In a game without a value we cannot predict deterministically the outcome of the game played by rational players. In fact, each such game generates a genuine intelligence conflict where each player tries to guess his opponent's strategic choice. This is a particular instance of the competition for the second move, analyzed in Chapter 2 Section 2. Our first example is a game with a value.

Example 1.   Noisy gunfight.

Both players have one bullet in their gun and walk toward each other at constant speed. At time $t = 0$ they are far apart, and at time $t = 1$ they collide. The real-valued function a_i, defined on $[0, 1]$, measures the skill of Player i, $i = 1, 2$. Namely, $a_i(t)$ is the probability that Player i hits player j if he shoots at time t. We assume that a_i is strictly increasing continuously and that $a_i(0) = 0$, $a_i(1) = 1$. The payoff is +1 if Player 1 hits Player 2 before being shot, -1 in the symmetrical case, and zero if no one has been hurt or both are hit simultaneously.

The strategy sets are $X_1 = X_2 = [0, 1]$. Strategy x_i for Player i means: I will shoot at time $t = x_i$ if the opponent has not shot yet. If he has and did not hit me, I will shoot safely at time $t = 1$. Thus, the normal form of the game is (X_1, X_2, u) where

$$u(x_1, x_2) = \begin{cases} 2\,a_1(x_1) - 1 & \text{if } x_1 < x_2 \\ a_1(x_1) - a_2(x_1) & \text{if } x_1 = x_2 \\ 1 - 2\,a_2(x_2) & \text{if } x_2 < x_1 \end{cases}$$

For instance, assume $x_1 = x_2$. Then the payoff is +1 with probability $a_1(x_1) \cdot (1 - a_2(x_1))$. That is if Player 1 hits his opponent and is not hit. It is -1 with probability $a_2(x_1) \cdot (1 - a_1(x_1))$. To check that the game has a value, we compute its prudent strategies. Take Player 1 and compute $\phi(x_1) = \inf_{0 \le x_2 \le 1} u(x_1, x_2)$:

$$\phi(x_1) = \inf \{2a_1(x_1) - 1, 1 - 2a_2(x_1)\}$$

Let x^* be defined by $2a_1(x^*) - 1 = 1 - 2a_2(x^*)$; i.e., $a_1(x^*) + a_2(x^*) = 1$. Then ϕ increases on $[0, x^*]$ (where it coincides with $2a_1 - 1$) and decreases on $[x^*, 1]$ (where it coincides with $1 - 2a_2$). Thus $\alpha_1 = 2a_1(x^*) - 1 = 1 - 2a_2(x^*)$, and Player 1's unique prudent strategy is x^*. A symmetrical computation shows that it is Player 2's prudent strategy as well and $\alpha_2 = \alpha_1$. Thus both optimally shoot at time x^*. Another way of proving the result would be to check directly that (x^*, x^*) is a saddle point of u. This might be shorter, but you need to guess both palyers' optimal strategies.

The simplest example of a game without a value is matching pennies. Player 1 hides either no penny or one penny in his hand; Player 2 tries to guess what he has done. The payoff is +1 if he guesses what Player 1 has done correctly and -1 otherwise. It is convenient to represent this game in matrix form. Here Player 1 picks a row and Player 2 a column. The corresponding entries show the payoff. It is apparent that $\alpha_1 = -1 < +1 = \alpha_2$.

$$
\begin{array}{c}
\text{Player 1}
\end{array}
\quad
\begin{array}{c}
1 \\
\\
0
\end{array}
\left[
\begin{array}{cc}
+1 & -1 \\
\\
-1 & +1
\end{array}
\right]
$$

$$
\begin{array}{cc}
0 & 1
\end{array}
$$

Player 2

Remark 1.

If the strategy sets X_i, $i = 1, 2$ are compact topological spaces and u is continuous, then each player has at least one prudent strategy. In general, a game might fail to have a prudent strategy but if u is uniformly bounded on $X_1 \times X_2$, each player has an ε prudent strategy for arbitrary small $\varepsilon > 0$. x^*_i is ε prudent if

$$
\inf_{x_2 \in X_2} u(x^*_1, x_2) \geq \alpha_1 - \varepsilon
$$

We need the assumption that u is uniformly bounded to guarantee that both α_1 and α_2 are finite. These facts imply

i) Suppose X_i, $i = 1, 2$ are compact and u is continuous.

18

Then, if G has a value, it has (at least) one saddle point.

ii) Suppose u is uniformly bounded on X_1 x X_2. Then if G has a value, for arbitrary small ε > 0, it has (at least) one pair of ε prudent strategies which in turn is a 2ε saddle point of G. Conversely, any ε saddle point of G is made up of 2ε prudent strategies. The precise statement of claim ii and its proof are the subject of Exercise 11.

2. VON NEUMANN'S THEOREM

Two powerful theorems give sufficient conditions for a game G = (X_1, X_2, u) to have a value. The first one relies upon the convexity properties of the strategy set and the payoff function. The second theorem is presented in the next section.

Theorem 2 (von *Neumann*).

Suppose X_i, i = 1, 2 are convex compact subsets of two topological vector spaces and u is continuous on X_1 x X_2. Suppose, moreover, for all x_2 ε X_2, u(·, x_2) is quasi concave in x_1; for all x_1 ε X_1, u(x_1, ·) is quasi convex in x_2. Then the game (X_1, X_2, u) has a value and (at least) one saddle point. Actually, Player i's prudent strategies form a convex compact subset of X_i, i = 1,2.

The result of theorem 2 will be proved as a corollary of Nash's theorem (Chapter 5, Section 4). Its principal

19

application is to the mixed extension of a two-person,
zero-sum Game (Chapter 7). For the time being, we propose
a simple example.

Example 2. Borel's model of poker.

Each player bids first $1. Next he receives a hand,
namely, a number $m_i \in [0, 1]$ i = 1, 2. Both hands are
uniformly distributed on $[0, 1]$ and independent. Each
player observes only his hand.

Player 1 moves first. He can fold (in which case,
Player 2 collects the initial bids). Or he can bid $5 more,
in which case Player 2 has to move. He can fold (Player 1
collects all bids) or bid $5 as well, in which case both
hands are uncovered. The player with the highest hand is
declared the winner and collects the pot.

In principle, a strategy by Player i is any mapping
from $[0, 1]$ into {fold, bid} making his choice (to bid or
to fold) depend upon his hand in an arbitrary way. It is
not necessary, however, to use this complicated strategy set
(a rigorous justification of this fact is the subject of
Exercise 9, Chapter 3). The following simple strategies
will do:

for i = 1, 2: $X_i = [0, 1]$, x_i means Player i folds

whenever $m_i \leq x_i$ and bids $5 otherwise

Notice that for Player 2, bidding is effective only if Player 1 also bids. Let us compute the payoff $u(x_1, x_2)$. With probability $(1 - x_1) \cdot (1 - x_2)$ both players bid and $u = \pm 6$ according to $m_1 > m_2$ or $m_1 < m_2$ ($m_1 = m_2$ occurs with probability zero). Moreover the probability $\pi(x_1, x_2)$ that $m_1 > m_2$ given that $x_i \leq m_i \leq 1$ is easily computed as:

$$\pi(x_1, x_2) = \frac{1 + x_1 - 2x_2}{2(1 - x_2)} \qquad \text{if } x_2 \leq x_1$$

$$\pi(x_1, x_2) = \frac{1 - x_2}{2(1 - x_1)} \qquad \text{if } x_1 \leq x_2$$

Hence the payoff function

$$u = -6x_1^2 + 5x_1x_2 + 5x_1 - 5x_2 \qquad \text{if } x_2 \leq x_1$$

$$u = 6x_2^2 - 7x_1x_2 + 5x_1 - 5x_2 \qquad \text{if } x_1 \leq x_2$$

All assumptions of von Neumann's theorem hold true. u is continuous, concave in x_1 (for fixed x_2, u is linear and then becomes a downward parabola with the same derivative at $x_1 = x_2$), and convex in x_2. Thus u has a saddle point. To compute it one solves the system of first-order conditions $\partial u/\partial x_i = 0$, $i = 1, 2$. Trying $x_1 = x_2$ or $x_1 > x_2$ leads to a contradiction. Thus $x_1 < x_2$ and we must solve

$$\frac{\partial u}{\partial x_1} = -7x_2 + 5 = 0$$

$$\frac{\partial u}{\partial x_2} = 12x_2 - 7x_1 - 5 = 0$$

Hence the unique saddle point

$$x^*_1 = (\tfrac{5}{7})^2 \approx .51, \ x^*_2 = \tfrac{5}{7} \approx .71$$

The value of the game is $u(x^*_1, x^*_2) = -(5/7)^2 \approx -.51$. Thus Player 2 earns on average 51 cents by deciding to bid if his hand is above .71, and only in this case.

3. KUHN'S THEOREM

Example 3. Gale's chomp game.

On the oriented 8 x 8 chessboard (formally described as {1, 2, ..., 8} x {1, 2, ..., 8}) Player 1 picks an entry (i, j) (1 ≤ i, j ≤ 8) thus blocking all entries northeast of (i, j) (i.e., all (i', j') s.t. i ≤ i', j ≤ j'). Next, Player 2 picks one possible entry and blocks all entries northeast of it, and so on, until one player must block the (1, 1) entry (because it is the only one still open), which causes him to lose the game.

This is an example of a (two-person, zero-sum) game in extensive form. Players take turns to move and do so relying upon perfect information; the game must terminate after finitely many moves have been made (in chomp, a play cannot take more than 64 moves).

Kuhn's theorem asserts that such games always have a value. In a game like chomp, where one player must win (no draw), there is one player who <u>can</u> win no matter what his opponent does. Indeed, the normal form representation of the game (to be described rigorously after Definition 2 below) takes the form

$$u: \quad X_1 \times X_2 \rightarrow \{-1, +1\}$$

Hence the value must be either +1 (in which case Player 1 has a winning strategy x^*_1 : $u(x^*_1, x_2) = +1$ for all $x_2 \in X_2$) or -1 (in which case Player 2 has a winning strategy x^*_2 : $u(x_1, x^*_2) = -1$ for all $x_1 \in X_1$).

In the game of chomp, it is easily seen that Player 1 wins by the following strategy. First block (2, 2), leaving Player 2 with the first row and the first column; next copy Player 2's choice. If he blocks (1, i), i ≥ 2, then Player 1 should block (i, 1); if he blocks (i, 1), i ≥ 2, Player 1 should block (1, i).

Now consider the variant of chomp played on a 8 x 9 chessboard with identical rules. Kuhn's theorem again shows that <u>some</u> player must have a winning strategy. But this cannot be Player 2, for Player 1 can block (8, 9) first, wait Player 2's move — say (i, j) — and then move as Player 2's optimal strategy suggests in reply to (i, j). Since (i, j) is a legal first move of Player 1, Player 2's

winning strategy allows him to win against that opening.
This is a contradiction! Incidentally, actual description
of Player 1's optimal strategy in the 8 x 9 chomp game is
an open problem (see Gale [1974]).

Definition 2. Two-person, zero-sum games in extensive form.
A finite game in extensive form with perfect information
is given by i) a tree (connected graph without cycles)
with a particular node taken as the origin; ii) for each
terminal node, a (real valued) payoff attached to it;
iii) for each nonterminal node, a specification of who
has the move.

Formally, a finite tree is a pair $\Gamma = (M, \sigma)$ where M is the
finite set of nodes, where σ associates to each node its
nearby predecessor. In addition, there is a unique node m_o,
such that $\sigma(m_o) = m_o$, which we call the origin of Γ; there is
an integer ℓ such that $\sigma^\ell(m) = m_o$ for all $m \in M$, the smallest
of which is the length of Γ.

A node m such that $\sigma^{-1}(m) = \phi$ is a terminal node of Γ,
and their set is denoted T(M). For a nonterminal node m,
$\sigma^{-1}(m)$ is the set of successors of m.

Given a tree Γ, a two-person, zero-sum game is defined
by a partition of M as $M = T(M) \cup M_1 \cup M_2$, and a payoff function
u defined over T(M).

The partition M_i, i = 1, 2 specifies which player has
the move at each particular nonterminal node. If $m_o \in M_i$,

then at the beginning of the play, Player i must pick a
successor of m_o, namely, a node m_1 in $\sigma^{-1}(m_o)$. If $m_1 \varepsilon$ T(M)
is a terminal node, the play is over and the corresponding
payoffs are $u(m_1)$. If $m_1 \notin$ T(M) is a nonterminal node, then
Player j, such that $m_1 \varepsilon M_j$, has the move and picks a
successor of m_1, namely, a node $m_2 \varepsilon \sigma^{-1}(m_1)$. And so on.

A game in extensive form (Γ, M_1, M_2, u) has a canonical
normal form (X_1, X_2, $\tilde{u}$) where a strategy of Player i specifies
his move at each and every node where he has to play.
Formally, X_i is the set of mappings x_i from M_i into M such
that $x_i(m) \varepsilon \sigma^{-1}(m)$, all $m \varepsilon M_i$. And $\tilde{u}(x_1, x_2) = u(m)$ where
m is the node at which the play generated by (x_1, x_2)
terminates.

Theorem 3 (Kuhn [1953]).

Every finite two-person, zero-sum game in extensive
form has a value. Also, each player has at least
one optimal (prudent) strategy.

Proof.
The proof is by induction on the length ℓ^* of the tree Γ.
For $\ell = 1$ the claim holds trivially since we are dealing with
one-person games. Suppose it holds for $\ell = 1, \ldots, L$ and
consider a tree Γ of length $L + 1$, and an associated game
(Γ, M_1, M_2, u). Without loss of generality, say that
Player 1 has the first move and denote $\sigma^{-1}(m_o) = \{m_1, \ldots, m_t\}$

the set of its successors. The subtree of Γ starting at
any node m_τ, $1 \le \tau \le t$, defines a game G_τ in extensive
form with length at most L. Here is an example.

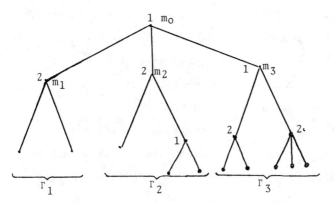

By the induction assumption, each game G_τ has a value,
say, v_τ. We claim that the value of our original game is
$v = \sup_{1 \le \tau \le t} v_\tau$. Indeed by moving first to m_τ and then playing
optimally in G_τ, Player 1 is guaranteed of the utility
level v. Also by playing optimally in each game $G_1, \ldots, G_t$
Player 2 guarantees a disutility level not above v. <u>QED</u>.

Applications of Kuhn's theorem are numerous. One
example is Zermelo's theorem. In chess either White can
force a win, or Black can force a win, or both can force a
draw (chess is a finite game because if a board position is
repeated three times we have a draw). A rich family of
games to which Kuhn's theorem applies are the Nim-type
games (Gale and Neymann [1980]) of which chomp is an example.
Following is another one.

Example 4. Splitting the pile.

We start with a pile of n tokens ($n \geq 2$). Initially Player 1 must split it into two piles of height n_1, n_2 (as a rule $n_i \geq 1$, $i = 1$, 2). If either one of n_1, n_2 is 1, Player 2 picks it and wins. Otherwise, Player 2 must pick one pile of height at least 2 (thus discarding the other pile) and split it into two nonempty piles, and so on, until one player can pick a singleton and win.

By Kuhn's theorem we can partition the integers {1, 2, ...} as $N_1 \cup N_2$ where N_i contains those n such that Player i can force a win if the initial pile has height n. Clearly $1 \in N_1$. Moreover,

$$\forall \, n \in N_1, \, n \geq 2 \; \exists \; n_1, \, 1 \leq n_1 \leq n - 1; \quad n_1 \in N_2 \text{ and } n - n_1 \in N_2$$

$$\forall \, n \in N_2, \, \forall \, n_1, \, 1 \leq n_1 \leq n - 1; \quad n_1 \in N_1 \text{ and/or } n - n_1 \in N_1$$

This system uniquely characterizes N_1 as the set of those integers who are 0, 1, or 4 modulo 5.

4. EXERCISES

a) Normal form games: examples

1) Gale's roulette

Each wheel has an equal probability to stop on any of its numbers. Player 1 chooses a wheel and spins it. While it is still spinning Player 2 chooses another wheel and

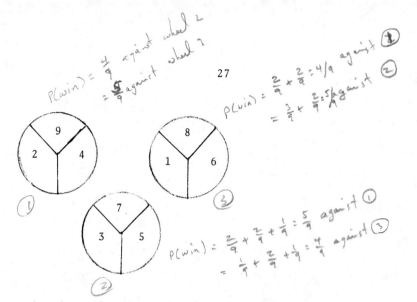

$P(win) = \frac{4}{9}$ against wheel 2
$= \frac{5}{9}$ against wheel 3

$P(win) = \frac{2}{9} + \frac{2}{9} = 4/9$ against ①
$= \frac{3}{9} + \frac{2}{9} = 5/9$ against ②

$P(win) = \frac{2}{9} + \frac{2}{9} + \frac{1}{9} = \frac{5}{9}$ against ①
$= \frac{1}{9} + \frac{2}{9} + \frac{2}{9} = \frac{4}{9}$ against ③

spins it. The winner is the player whose wheel stops on
the bigger score. He receives $1 from the loser.

Which player would you like to be in this game? Player 2

If you are the winning player, at what price would
you sell your seat in the game to a third party (assuming
you are risk neutral and compare random events by their
expected payoff)?

2) <u>Land division game</u>

The land consists of three contiguous pieces: a square
with corner $(0, 0),(1, 0),(1, 1),(0, 1)$; a triangle with
corners $(0, 1),(1, 1),(0, 2)$; and a triangle with corners
$(1, 0),(1, 1),(2, 1)$. The method of division is as follows.
Simultaneously Player 1 chooses a vertical line L_v with the
x coordinate in $[0,1]$, and Player 2 chooses a horizontal
line L_h with the y coordinate in $[0,1]$. Then Player 1 gets
all the land which is above L_h and to the left of L_v as
well as all the land which is below L_h and to the right of

L_v. Player 2 gets the rest. Find the value of the game and optimal strategies, assuming both players want as much land as possible.

3) Silent gunfight

This is a variant of Example 1. However, here a shot is fired which cannot be heard. Thus you are not aware that your opponent has shot unless you are hit. The payoff function is now:

$$u(x_1, x_2) = \begin{cases} a_1(x_1) - a_2(x_2) + a_1(x_1) \cdot a_2(x_2) & \text{if } x_1 < x_2 \\ a_1(x_1) - a_2(x_2) & \text{if } x_1 = x_2 \\ a_1(x_1) - a_2(x_2) - a_1(x_1) \cdot a_2(x_2) & \text{if } x_2 < x_1 \end{cases}$$

Justify this formula and prove that this game has no value and that the value v of the noisy duel is such that $\alpha_1 < v < \alpha_2$.

Hint: Compute $\alpha_1 = \sup_{x_1} \inf \{2a_1(x_1) - 1, a_1(x_1) - a_2(x_1) - a_1(x_1) \cdot a_2(x_1)\}$. Defining $\tilde{t}$ by $a_1(\tilde{t}) + a_1(\tilde{t})a_2(\tilde{t}) = 1$, this can be rewritten as

$$\alpha_1 = \sup_{x \geq \tilde{t}} \{a_1(x) - a_2(x) - a_1(x)a_2(x)\}$$

Similarly

$$\alpha_2 = \inf_{x \geq \tilde{t}} \{a_1(x) - a_2(x) + a_1(x)a_2(x)\}$$

Assume first that $a_1 - a_2 - a_1 a_2$ is nonincreasing and
$a_1 - a_2 + a_1 a_2$ is nondecreasing.

4) Borel's model of poker continued

Generalize the computation of Example 2 where the second
bid of Player 1 is worth \$a, a $\geq$ 1, and that of Player 2 is
worth \$b, b $\geq$ 1. Discuss according to the sign of
$a^2 + 2a - 2b$.

5) Final offer arbitration (Brams and Merrill [1983])

Player 2 must pay \$x to Player 1 where x is determined as
follows. Each player proposes an amount x_i, i = 1, 2. Next,
the referee picks whichever of x_1, x_2 is closest to his own
opinion y. Both players are aware of the probability distribution from which y is drawn. This distribution has a positive
continuous density f on $]-\infty, +\infty[$. Denoting by F the cumulative
distribution of f, the following game is in order.

$$X_1 = X_2 =]-\infty, +\infty[$$

$$u(x_1, x_2) = x_1 \cdot F\left(\frac{x_1 + x_2}{2}\right) + x_2 \cdot \left(1 - F\left(\frac{x_1 + x_2}{2}\right)\right) \text{ if } x_1 \leq x_2$$

$$= x_2 \cdot F\left(\frac{x_1 + x_2}{2}\right) + x_1 \cdot \left(1 - F\left(\frac{x_1 + x_2}{2}\right)\right) \text{ if } x_2 \leq x_1$$

Solve the system of first-order conditions for a saddle
point of u and show that it has two solutions (α, β) and
(β, α). Prove that at a saddle point of u we must have

$x_2 \leq x_1$, which leaves us with only one possible saddle point. Write the second-order conditions: $\partial^2 u/\partial x_1{}^2 \leq 0$ and $\partial^2 u/\partial x_2{}^2 \geq 0$, assuming f is derivable. Conclude by giving an example of f where our game has a value and one where it does not.

b) Games in extensive form: examples

6) Splitting the pile: continued

Solve the variant of Example 4 where the player who must pick a singleton loses.

7) Removing sticks

Players successively remove one stick or two adjacent sticks. The initial position is

For instance, if Player 1 removes first 5, 6, Player 2 has 5 legal moves: remove 1 or 3 or 4 or 3, 4 or 7.
Who wins if the player removing the last stick loses? Who wins if the player removing the last stick wins?

8) The Marienbad game

Physically the game amounts to a set of p rows of respectively n_1, ..., n_p sticks. Each player is asked to remove successively at least one and at most all sticks in exactly one row still alive. The player who picks the last stick loses.

We fix an integer p, p ≥ 1 and for all p-tuples $\underline{n}$ = $(n_1, \ldots, n_p)$ where $n_1, \ldots, n_p$ are integers, possibly zero, we define two games, $G^1_{\underline{n}}$, $G^2_{\underline{n}}$, as follows. Given $\underline{n} \neq (0, 0, \ldots, 0)$, call $\underline{n}'$ a <u>successor</u> of $\underline{n}$ if there exists k, $1 \leq k \leq p$ such that

$$n'_{k'} = n_{k'} \quad \text{for all } k', \ 1 \leq k' \leq p \quad \text{and} \quad k' \neq k$$

$$0 \leq n'_k < n_k$$

If $\underline{n}$ = $(0, \ldots, 0)$, then in $G^i_{\underline{n}}$ Player i wins. If $\underline{n} \neq (0, \ldots, 0)$, then $G^i_{\underline{n}}$ is played as follows. First, Player i picks a successor $\underline{n}'$ of $\underline{n}$. If $\underline{n}'$ = $(0, \ldots, 0)$, then Player j, $j \neq i$, wins the play.

Otherwise, the game $G^j_{\underline{n}}$ starts. Player j picks a successor $\underline{n}''$ of $\underline{n}'$, if $\underline{n}''$ = $(0, \ldots, 0)$ Player i wins. Otherwise $G^i_{\underline{n}''}$ starts. And so on.

a) Prove the existence of a partition, $N_1 \cup N_2$ of $\{0, 1, 2, \ldots\}^p$ such that if $\underline{n}$ belongs to N_i, Player i can force a win in $G^1_{\underline{n}}$.

b) For all $\underline{n}$ = $(n_1, \ldots, n_p)$ denote

$$n_k = \alpha^k_\ell \alpha^k_{\ell-1}, \ldots, \alpha^k_1 \alpha^k_0, \qquad 1 \leq k \leq p$$

the binary representation of n_k where ℓ is an upper bound of the number of digits needed for $n_1, \ldots, n_p$. (Thus some, not all, among $\alpha_\ell^1, \ldots, \alpha_\ell^p$ might be zero.) Next denote

$$a_o = \sum_{k=1}^{p} \alpha_o^k, \quad \ldots, \quad a_\ell = \sum_{k=1}^{p} \alpha_\ell^k$$

Then prove that $N_2 = M_2 \cup P_2$ where

$M_2 = \{\underline{n}/ \; \forall \; j, \; 0 \leq j \leq \ell, \; a_j \text{ is even and } \exists j, \; 1 \leq j \leq \ell, \; a_j > 0\}$

$P_2 = \{\underline{n}/ \; \forall \; j, \; 1 \leq j \leq \ell, \; a_j = 0 \text{ and } a_o \text{ is odd}\}$

c) <u>Abstract normal form games</u>

9) <u>Symmetrical games</u>

A two-person, zero-sum game is symmetrical if $X_1 = X_2 = X$ and moreover

$$u(x_1, x_2) = -u(x_2, x_1) \qquad \text{all } x_1, x_2 \in X$$

Prove that the value of a symmetrical game (if any) is zero and optimal strategies of both player coincide. What if the game has no value?

10) <u>Shapley's criterion for finite games</u>

Let $G = (X_1, X_2, u)$ be a finite game (both X_1, X_2 are finite) such that for all doubletons $Y_1 \subset X_1, Y_2 \subset X_2$ the restricted game (Y_1, Y_2, u) has a value. Show that G has a value.

Hint: Assume G has no value.

Without loss, assume $\alpha_1 < 0 < \alpha_2$. Pick a strategy $x_2^* \in X_2$ which maximizes the integer-valued function ρ:

$$\rho(x_2) = \divideontimes \{x_1 \in X_1 | u(x_1, x_2) \leq 0\}$$

One can find an $x_1^* \in X_1$ such that $u(x_1^*, x_2^*) > 0$. (Why?) Next one can find $\tilde{x}_2$ such that $u(x_1^*, \tilde{x}_2) < 0$ (why?).

11) Proof of Remark 1

Let $G = (X_1, X_2, u)$ be a two-person, zero-sum game where X_i, $i = 1, 2$ are arbitrary sets. Define x_1^* to be an ε-prudent strategy of Player 1 iff $u(x_1^*, x_2) \geq \alpha_1 - \varepsilon$ for all x_2 and x_2^* to be an ε-prudent strategy of Player 2 iff $u(x_1, x_2^*) \leq \alpha_2 + \varepsilon$ for all x_1. Similarly, define an ε saddle point by $u(x_1, x_2^*) - \varepsilon \leq u(x_1^*, x_2^*) \leq u(x_1^*, x_2) + \varepsilon$, for all x_1, x_2. Throughout the exercise we assume that u is uniformly bounded on $X_1 \times X_2$. There exist two real numbers a, b such that

$$a \leq u(x_1, x_2) \leq b \quad \text{all } (x_1, x_2) \in X_1 \times X_2$$

Prove the following claims. a) For any $\varepsilon > 0$, each player has an ε-prudent strategy in G. b) If G has a value and $x^* = (x_1^*, x_2^*)$ is a pair of ε-prudent strategies, then x^* is a 2ε saddle point. c) If x^* is an ε saddle point, then x_i^* is an 2ε prudent strategy of Player i (for i = 1, 2) and $(\alpha_2 - \alpha_1) \leq 2\varepsilon$. d) If, for all $\varepsilon > 0$, G has an ε saddle point, then G has a value.

12) A topological duel (Choquet)

Let E be a metric space. We denote by 0 the set of subsets of E with a nonempty interior. Our game works as follows:

In Step 1, Player 1 picks an $A_1 \in 0$. In Step 2, Player 2 picks an $A_2 \in 0$ with the only constraint $A_2 \subset A_1$.... In Step t a player (1 for odd t, 2 for even t) picks an $A_t \in 0$ with the only constraint $A_t \subset A_{t-1}$ and so on undefinitely. We say that Player 1 wins the play if

$$\bigcap_{t=1}^{\infty} A_t \neq \emptyset$$

If this intersection is empty, we say that Player 2 wins the play. a) Prove that if E is a complete metric space, Player 1 can force a win. b) Prove that if E = Q (the rational numbers), Player 2 can force a win. c) Are the results affected if both players must now pick nonempty open subsets of E?

CHAPTER 2. TACTICAL EXCHANGES OF INFORMATION

[handwritten margin notes: n-person, constant-sum games in which each player has an "optimal" stgy.]

[handwritten: i.e. a strategy x_i^* s.t., no matter what actions the other players select, player i should select x_i^*. That is, the BCA do no better than to play x_i^*]

1. INESSENTIAL GAMES

Inessential games are the (small) class of n-person games where each player has an optimal strategy. They generalize two-person, zero-sum games with a value.

Notation. Given the strategy sets X_1, ..., X_n, we denote X_N the cartesian product $X_1 \times \ldots \times X_n$. For any i, $1 \leq i \leq n$, we denote X_{-i} the cartesian product of the X_j for all $j \neq i$, $1 \leq j \leq n$. Its current element is denoted x_{-i}.

Definition 1.

In the n-person normal form game $G = (X_i, u_i, i = 1, \ldots, n)$, the *secure* utility level to player i is defined by

$$\alpha_i = \sup_{x_i \in X_i} \inf_{x_{-i} \in X_{-i}} u_i(x_i, x_{-i})$$

We say that $x^*_i \in X_i$ is a _prudent_ strategy of player i, $1 \le i \le n$, iff

$$\inf_{x_{-i} \in X_{-i}} u_i(x^*_i, x_{-i}) = \alpha_i$$

Definition 2.

The game G is _inessential_ if the utility vector $(\alpha_1, \ldots, \alpha_n)$ is Pareto undominated. Namely,

$$\text{for } \underline{no} \text{ outcome } y \in X_N \begin{cases} \alpha_i \le u_i(y) & \text{all } i = 1, \ldots, n \\ \\ \alpha_i < u_i(y) & \text{at least one } i \end{cases}$$

The intuition behind inessential games is as follows. One unit of an homogeneous cake is to be divided among our n players. Suppose Player i — by playing well — can guarantee that his final share is at least α_i, whatever is the behavior of other players. Suppose, moreover, that $\sum_{i \in N} \alpha_i = 1$. Then $(\alpha_1, \ldots, \alpha_n)$ is the sharing of the cake that must result from the "optimal" behavior by the players. In our general framework, utility levels are not comparable between players. Hence, they cannot be added. Yet in an inessential game, prudent strategies are optimal in the following sense.

Theorem 1.

Let $(X_i, u_i; i = 1, \ldots, n)$ be an _inessential_ game. For all $i = 1, \ldots, n$ let x^*_i be a _prudent_ strategy

*of i, and let x^* be the associated outcome. Then,
1) $u_i(x^*) = \alpha_i \le u_i(x^*_i, y_{-i})$ for all $i = 1, \ldots, n$ and
$y_{-i} \in X_{-i}$ 2) x^* is Pareto optimal 3) for all coalitions
$S \subset N$ and all strategy S-tuples $y_S \in X_S$, the following
conditions together are impossible.*

$$\forall \; i \in S \quad u_i(x^*) \le u_i(y_S, x^*_{S^c})$$

$$\exists \; i \in S \quad u_i(x^*) < u_i(y_S, x^*_{S^c})$$

Proof.

Since x_i is a prudent strategy of agent i, we have

$$\alpha_i = \inf_{y_{-i} \in X_{-i}} u_i(x^*_i, y_{-i}) \le u_i(x^*)$$

This inequality holds for all i. As our game is inessential,
this implies $\alpha_i = u_i(x*)$ for all i. Hence Property 1.

Property 2 follows from Property 3 by making S = N. To
prove 3 we pick S and $y_S \in X_S$ such that

$$\forall \; i \in S \; \alpha_i \le u_i(y_S, x^*_{S^c}) \tag{1}$$

Applying property 1 to $j \in S^c$ we get

$$\forall \; j \in S^c \; \alpha_j \le u_j(y_S, x^*_{S^c})$$

Combining these two systems of inequalities yields
$u_i(y_S, x^*_{S^c}) = \alpha_i$ for all i, because our game is inessential.
Thus, no inequality can be strict in (1). QED.

Note: these are constant-sum games

Property 1 states that if Player i uses an optimal
(i.e., prudent) strategy and expects the other players to
do the same, he enjoys the utility level α_i; should some
players j, j $\neq$ i, fail to use an optimal strategy, this can
only be profitable to Player i.

Property 3 means that neither a single player nor a
coalition of players has any incentive to deviate from prudent
strategies unilaterally (we implicitly assume that no side
payments can be made within a coalition because utilities are
not transferable). In terms of Definition 1, Chapter 10, we
say that an n-tuple of prudent strategies is a strong
equilibrium. Property 2 is just a particular case of the
above.

The main example of inessential games are zero-sum games
with a value.

Lemma 1.

*Let $G = (X_1, X_2, u, -u)$ be a two-person, zero-sum game.
If G has a value, it is inessential. Conversely, if G is
inessential, and each player has at least one prudent strategy,
then G has a value.*

The straightforward proof is omitted.

Remark 1.

This remark parallels Remark 1, Chapter 1. Without additional
topological assumptions (e.g., X_i is compact and u_i is

continuous on X_N, all i = 1, ..., n) existence of a prudent strategy (whether the game is inessential or not) is not guaranteed. However, assuming u_i is uniformly bounded on X_N for all i, we know that for arbitrarily small ε, each player has an ε prudent strategy. If G is inessential, a n-tuple x^* of ε prudent strategies will satisfy Properties 1, 2, 3 of Theorem 1 witn an error of at most ε. For instance x^* is ε Pareto optimal means that no outcome y exists such that $u_i(y) \geq u_i(x^*) + \varepsilon$ for all i = 1, ..., n. The proof of these claims is the same as that for Exercise 11, Chapter 1.

To interpret Definition 2 further, let us consider a noninessential game G. Then for every outcome x ε X_N of G, at least one of the following holds, barring optimality of x. a) Some Player i has $u_i(x) < \alpha_i$; he would guarantee a better utility by using a prudent strategy (or an ε prudent one for ε small enough). b) Some Player i has $u_i(x) > \alpha_i$, so that the utility $u_i(x)$ is not guaranteed to i. It follows that i can be threatened by the other players, who have the strategic power to lower i's utility. This vulnerability to threats jeopardizes the stability of x (this phenomenon is analyzed in Section 3 below). c) The utility vector $(u_1(x), ..., u_n(x))$ fails to be Pareto optimal; thus (global) cooperative arguments exist to depart (jointly) from outcome x.

2. COMPETITION FOR THE SECOND MOVE

"If the enemy thinks of the mountains, attack like
the sea; and if he thinks of the sea, attack like the
mountains" (Musashi [1981]). Warfare games ("kriegspiel")
offer beautiful examples of competition for the second move.
An attacker must choose where to land his troops, while the
defender allocates his forces among the feasible locations.
Whoever outguesses his opponent's strategic choice wins.
The attack fails if the defender concentrates his forces at
the very landing place, but no matter how he allocates his
forces, there is always at least one location where an attack
would succeed.

In such a game you want to spy out your opponent without
being spied out yourself. All information is considered to
be unreliable since your opponent may try to mislead you;
hence the paranoia of persecution familiar to intelligence
games.

Let (X_1, X_2, u) be a two-person, zero-sum game without
a value:

$$\sup_{x_1} \inf_{x_2} u(x_1, x_2) = \alpha_1 < \alpha_2 = \inf_{x_2} \sup_{x_1} u(x_1, x_2)$$

Say that Player 1 (respectively Player 2) wins by forcing
the payoff above α_2 (respectively below α_1).

Start with Player 1 considering to play x^o_1. If he anticipates this choice, Player 2 wins by using a best reply x^1_2 :

$$u(x^o_1, x^1_2) = \inf_{y_2} u(x^o_1, y_2) \leq \alpha_1$$

If Player 1 anticipates x^1_2 (e.g., his move toward x^o_1 was a fake), he wins by any best reply x^1_1 :

$$\alpha_2 \leq \sup_{y_1} u(y_1, x^1_2) = u(x^1_1, x^1_2)$$

Of course Player 2 only pretended to believe Player 1's initial message, so he now wins by replying x^2_2 :

$$u(x^1_1, x^2_2) = \inf_{y_2} u(x^1_1, y_2) \leq \alpha_1$$

And so on.

Two sequences (x^t_1, x^t_2) of mutual expectations arise where

$$u(x^{t-1}_1, x^t_2) \leq \alpha_1 < \alpha_2 \leq u(x^t_1, x^t_2) \tag{2}$$

for all $t = 1, 2, \ldots$ Also, each player tries to remain one step (not two!) ahead of the other. In fact, inequalities (2) imply that neither x^t_1 nor x^t_2 can be a converging sequence when t goes to infinity (as an exercise, assume X_1, X_2 are compact, u is continuous, and prove this claim).

We define the competition for the second move in an arbitrary two-person game — not necessarily zero-sum.

Definition 3.

Let $G = (X_1, X_2, u_1, u_2)$ be a two-person game.
Denote by $\beta_i = \inf\limits_{x_j} \sup\limits_{x_i} u_i(x_i, x_j)$ the guaranteed
utility level to i playing second. We say that
the competition for the second move arises in G
if the utility vector (β_1, β_2) is not feasible.
Thus, there is no outcome $x \in X_{12}$ such that
$\beta_i \leq u_i(x)$ $i = 1, 2$.

Player 1 has the second move (plays second), which means that
he observes Player 2's final choice x_2 before making any
decision about his own strategy x_1. He is <u>not</u> allowed to
take any commitment about x_1; he <u>must</u> listen to the announce-
ment of x_2 before choosing x_1. By optimally replying to x_2,
Player 1 derives the utility $\sup\limits_{x_1} u_1(x_1, x_2)$, hence $\beta_1 = $
$\inf\limits_{x_2} \sup\limits_{x_1} u_1$ is guaranteed to him if he plays second. If the
competition for the second move arises in G, no matter what
outcome x we consider, at least one player would be better
off by taking the second move.

A zero-sum game displays the competition for the second
move if it has no value. An example of this competition in
nonzero-sum games is proposed in Exercise 5.

3. COMPETITION FOR THE FIRST MOVE

Competition for the first move means that a player wins
by playing first, i.e., committing himself to use a particular

strategy before any other player is committed. Bargaining
for a price between a buyer and a seller (or for a salary
between an employer and the employees) is a conflict of
this kind: I want to persuade my opponent that my offer is
"take it or leave it" while denying that my opponent's
last offer was. That offer which is finally accepted by
the receiving party as irrevocable wins.

As in the competition for the second move, the critical
variable is the "true" level of commitment. There, each
player was delaying as much as possible his final strategic
decision so as to force his opponent to commit himself first
and hopefully be spied out. Here, on the contrary, each
tries to make the first commitment and denies that the
opponent might have.

Example 1. The crossing game.

Our players drive on two orthogonal roads and reach the
crossing simultaneously.

Each player can stop or go. The following 2 x 2 game
formalizes the situation by assuming that each player prefers
to stop rather than risk suffering an accident by continuing
(outcome (go, go)), and each prefers to go if the other stops.
The small number ε is the utility loss or gain that results
from viewing the other passing while one is safely stopping;
it varies according to the cultural pattern.

44

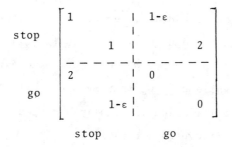

$$
\begin{array}{c}
\text{stop} \\[30pt]
\text{go}
\end{array}
\left[
\begin{array}{c|c}
1 \qquad\quad 1\text{-}\varepsilon \\
\quad\ 1 \qquad\qquad\quad 2 \\
\hline
2 \qquad\ \ 0 \\
\quad\ 1\text{-}\varepsilon \qquad\qquad\ 0
\end{array}
\right]
$$

stop   go

By committing himself to the nonprudent strategy of
continuing to go, a player wins, since he forces the other
to stop and therefore enjoys the maximal utility level 2.
Since no outcome offers the utility level 2 to <u>both</u> players,
the competition for the first move arises. Each player will
pretend that he has lost his ability to switch from go to
stop (e.g., by looking drunk) while at the same time he closely
observes his opponent to check whether or not he <u>really</u> cannot
stop anymore. The striking feature of these tactical moves
is that a skillful irrationality in one's behavior results in
winning, which turns out to be quite rational behavior after
all. The symmetry of both players' roles makes it impossible
to arbitrate the competition for the first move by means of
the sole normal form of the game.

Example 2. <u>War of attrition: the auction dollar game</u> (Maynard
Smith [1974]).

A prize will go to whoever endures a painful — or costly —
action the most. Think of the nuptial ritual of some bees,

during which males must fly vertically until all but one fall, or price wars where prices are cut so low that all firms lose money until the most patient captures the market and restores an acceptable profit margin. Formally, these situations are described by an auction game: two players bid x_i, with the only constraint $x_i \geq 0$. The highest bidder gets the \$1 prize; both players pay the lowest bid.

$$X_1 = X_2 = [0, +\infty[$$

$$u_1(x_1, x_2) = 1 - x_2 \qquad \text{if } x_2 < x_1$$

$$= \tfrac{1}{2} - x_1 \qquad \text{if } x_1 = x_2$$

$$= -x_1 \qquad \text{if } x_1 < x_2$$

$$u_2(x_1, x_2) = u_1(x_2, x_1)$$

The game is symmetrical (if you have to play this game, you do not mind being Player 1 or Player 2). The decision to bid \$1 or more ($x_i \geq 1$) wins. Realizing he can no longer make any profit, and would even lose money by bidding any positive amount, the other player abstains ($x_j = 0$), and the bidder gets the dollar at no cost!

Going now to a general two-person game, $G = (X_1, X_2, u_1, u_2)$, we compute the utility levels S_1, S_2 that each player expects when playing first in G. Say that Player 1 is the leader. He picks a strategy $x_1 \in X_1$, announces it to Player 2 who then chooses strategy x_2 among the possible best replies. Define BR_2 as the set of outcomes where Player 2 replies optimally to Player 1's strategy:

$$BR_2 = \{(x_1, x_2) \mid u_2(x_1, x_2) = \sup_{y_2 \in X_2} u_2(x_1, y_2)\}$$

As a leader, Player 1 may get at most the following utility level, called Player 1's Stackelberg utility level:

$$S_1 = \sup_{x \in BR_2} u_1(x) \tag{3}$$

Of course, Player 2 may have several best replies to a given strategy x_1 of Player 1, which are not indifferent to Player 1. In this event our definition assumes optimistically that Player 2's choice is altruistic. The opposite assumption leads to a smaller utility level S_i (see Exercise 4).

Similarly the best reply set BR_1 of Player 1 is

$$BR_1 = \{(x_1, x_2) \mid u_1(x_1, x_2) = \sup_{y_1 \in X_1} u_1(y_1, x_2)\}$$

Player 2 as a leader gets at most a Stackelberg utility level of S_2:

$$S_2 = \sup_{x \in BR_1} u_2(x) \tag{4}$$

A solution x^* of problem (3) ((4)) is called a 1-Stackelberg equilibrium of G (a 2-Stackelberg equilibrium):

$$x^* \in BR_2 \quad \text{and} \quad u_1(x^*) = S_1$$

(respectively $x^* \in BR_1$ and $u_2(x^*) = S_2$)

As for prudent strategies, Stackelberg equilibria exist when X_1, X_2 are compact and u_1, u_2 are continuous (e.g., X_1, X_2 are finite).

Definition 4.

In the two-person game G, we say that the competition for the first move arises if the utility vector (S_1, S_2) is not feasible:

There is no outcome $x \in X_{12}$ such that $S_i \le u_i(x)$ $i = 1, 2$

Under the competition for the first move, no matter what outcome is at stake, at least one player would be better off by taking the leadership. The terminology in Definitions 3 and 4 is consistent only if both types of competition cannot arise simultaneously.

Lemma 2.

Suppose both kinds of Stackelberg equilibria exist in game G. Then the competition for the first move and for the second move cannot happen together.

Proof.

Let x_1^i, $i = 1, 2$ be an i-Stackelberg equilibrium. From $x^1 \in BR_2$ follows

$$u_2(x^1) = \sup_{y_2} u_2(x_1^1, y_2) \geq \inf_{y_1} \sup_{y_2} u_2(y_1, y_2) = \beta_2 \qquad (5)$$

Similarly $x^2 \in BR_1$ implies $u_1(x^2) \geq \beta_1$.

If the competition for the second move arises, this implies $u_2(x^2) < \beta_2$. Then, by definition of x^2, $u_2(x^2) = S_2$. So $S_2 < \beta_2$. Next, by definition of x^1, $u_1(x^1) = S_1$. If the competition for the first move arises, this implies $u_2(x^1) < S_2$. Combined with (5) this gives $\beta_2 < S_2$, a contradiction. QED. (More examples of the competition for the first move are in Exercises 4 and 5.)

Definitions 3 and 4 first were given by Moulin [1979], although the underlying ideas were present in Luce and Raiffa [1957] and developed by Schelling [1971].

4. DETERRING THREATS

Example 3. Trading an indivisible object.

Player 1 (seller) selects a price $x_1 \in [0, 1]$ (conventionally zero is the seller's reservation price and one is the buyer's reservation price). Player 2 (buyer) buys or walks away: $x_2 \in \{0, 1\}$

$$u_i(x_1, 0) = 0 \qquad\qquad i = 1, 2$$

$$u_1(x_1, 1) = x_1 \qquad\qquad \text{all } x_1 \; \varepsilon \; [0, 1]$$

$$u_2(x_1, 1) = 1 - x_1$$

The seller wins by committing himself to a price close to one, which the buyer accepts because it is still his best reply. However, the buyer can force a price close to zero by threatening: I shall buy if $x_1 \leq .1$ or walk away. This seemingly irrational behavior (where the buyer refuses a still-profitable deal) proves to be quite rational if the threat is accepted by the buyer and he agrees to buy on the seller's terms. Using threats and commitment strategies as described above, we could end up anywhere between zero and one; alternatively, no trade takes place if both make threats which are not consistent (I sell for no less than .7; I do not buy above .4) and stick to them.

The point of Example 3 is that the policy of sharing the surplus available above the vector of secure utility level (zero to both in the example) can go wild if a player successfully threatens the other(s). Hence deterring threats yield great indeterminacy in the outcome of the game. This point can be given a rigorous formulation in all non-inessential two-person games.

In $G = (X_1, X_2, u_1, u_2)$ assume that Player 1 keeps all the cooperative surplus (above the secure utility level) at the outcome(s) $\tilde{x}(1)$ defined by

$$u_1(\tilde{x}(1)) = \sup_{u_2(x) \geq \alpha_2} u_1(x)$$

$$u_2(\tilde{x}(1)) \geq \alpha_2$$

A symmetrical definition holds for the outcome(s) $x(2)$ where Player 2 keeps all the surplus.

Lemma 3.   The wolf-sheep lemma.

In a two-person game, Player i keeps all the cooperative surplus if he successfully threatens Player j as follows. i) I will use $\tilde{x}_i$ (i) if you choose $\tilde{x}_j$ (i). ii) If you choose any other x_j, I will pick some x_i such that $u_j(x_i, x_j) = \inf_{y_i} u_j(y_i, x_j)$.

A wolf's threat requires that he deny any privacy to the sheep and make credible some reply strategies that might be very harmful to the wolf. The latter is essential. When the buyer announces that he will walk away if the price exceeds .1, his message is purely deterring, or so he hopes. Yet for this threat to be convincing, he must be ready to carry it out (or fake very well this determination). In this sense every deterring threat is risky. Think of the Indian bards who used to carry large sums of money and were protected only by the threat to commit suicide if robbed.

ıg [1971] Chapter 5, footnote 7.)

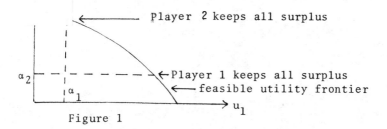

Figure 1

Conclusion

In inessential games neither the timing of the game
(who plays first) nor the informational context (who knows
what about other players preferences and/or strategical
choices) need to be specified. A clear subset of optimal
(prudent) strategies determine the rational outcome of the
game.

In non-inessential games (i.e., virtually all games
arising from economic and/or political models) these issues
matter a lot.

As optimal strategies are no longer available, the
rational behavior of our players (by which we mean that they
do their best to maximize their own utility, and their
intellectual abilities are unlimited) does not determine
in itself the outcome of the game. The bare normal form of
the game generates a variety of tactical exchanges of
information: threats, promises, commitment tactics,
spying, and lying, and so on. To sort out a few plausible

outcomes, we need additional assumptions about the information
and communication channels available to the players. These
assumptions form the basis of the analysis of Parts II and III.

5. EXERCISES

1) Give an example of a two-person game which is <u>not</u>
inessential but with a Pareto-optimal outcome x such that
$\beta_i = u_i(x)$, i = 1, 2.
<u>Note</u>: Outcome x is Pareto optimal if there is no outcome
y such that $u_i(x) \leq u_i(y)$, i = 1, 2 and at least one
inequality is strict).

2) If the competition for the second move arises in game
(X_1, X_2, u_1, u_2), prove that neither game $(X_1, X_2, u_1, -u_1)$
nor $(X_1, X_2, -u_2, u_2)$ has a value.

3) Give an example of a 2 x 2 game where an outcome
$x^* = (x_1^*, x_2^*)$ is a saddle point of u_1 and of $-u_2$:

$$u_1(x_1, x_2^*) \leq u_1(x_1^*, x_2^*) \leq u_1(x_1^*, x_2)$$

$$u_2(x_1^*, x_2) \leq u_2(x_1^*, x_2^*) \leq u_2(x_1, x_2^*)$$

yet the game is not inessential.

4) Both players pick an integer x_i, $1 \leq x_i \leq 10$. If
$x_1 + x_2 = 10$ then Player i's payoff is x_i. Otherwise, the
payoff is (4, 0) if $x_1 + x_2$ is even and (0, 4) if $x_1 + x_2$
is odd.

Show that the competition for the first move arises in this game. However, replacing the altruistic assumption by the opposite assumption in the definition of the utility levels S_i (so that a player, when indifferent towards choosing two best-reply strategies, chooses so as to minimize his opponent's utility) is enough to avoid this competition.

5) Strategic voting: Borda versus Rawls

There are seven candidates a, b, c, d, e, f, g and two voters. A ballot of Voter i, i = 1, 2 is a linear ordering such as

$$x_i = bdgecaf$$

$$\uparrow \qquad \uparrow$$

top   bottom

|| ||

best preferred   least preferred

Given two ballots (x_1, x_2) the elected candidate would turn out to be as follows.

Borda

Give six points for the top candidate in any ballot, five points to the second best and so on. Then add both scores to find a candidate's Borda score. Break ties by lexicographic ordering. For instance,

x_1	cgdebaf			abcdefg
x_2	bafgdce	Borda Score		6876348

Hence b is elected. We denote by $B(x_1, x_2)$ this voting rule.

Rawls

Define the score of a candidate as the smallest of the two scores. Again break ties lexicographically. In the example above

	abcdefg
Rawls Score	1212003

g is elected. Denote $R(x_1, x_2)$ for this voting rule.

Voters cast their ballots strategically, hence they are not necessarily truthful. Let u_i be Voter i's true pre-ference, written as a fixed-scale utility function with range $\{0, \ldots, 6\}$ (hence voters are never indifferent about any two candidates).

The two voting rules described above generate two games

$$(X_1, X_2, u_1B, u_2B)$$

$$(X_1, X_2, u_1R, u_2R)$$

where $X_1 = X_2$ is the set (with cardinality 7!) of all linear orderings of our seven candidates.

a) Show that

$$\sup_{x_i} \inf_{x_j} u_i B(x) \leq 1 \leq 5 \leq \inf_{x_j} \sup_{x_i} u_i B(x).$$

Deduce that, in general, competition for the second move

arises in the Borda game. Make clear when it does not arise.

b) Show that

$$\sup_{x_i} \inf_{x_j} u_i R(x) = \inf_{x_j} \sup_{x_i} u_i R(x) = 3$$

and deduce that the competition for the second move never arise in the Rawls game. Next show that Player i's optimal utility as a leader is

$$S_i = \sup \{u_i(z) | u_j(z) \geq 3\}$$

and deduce that in general the competition for the first move arises in the Rawls game (again, give the exact meaning of "in general").

Hint: This is rather difficult as it involves a two-stage modelling: first, the voting rules B and R to derive an elected candidate from a pair (x_1, x_2) of arbitrary ballots, and, second, the game resulting from the specification of our players preferences. Proceed step by step. In a), for instance, choose an arbitrary u_1 and think about Player 1 trying to cast a ballot that effectively eliminates his least-acceptable candidate. Is that possible? Next has he any way to eliminate the candidate who is preferred next to last? Think of Player 2 knowing Player 1's ballot and putting all his skill to enforce election of the candidate who is preferred next to last by Player 1.

PART II.   NONCOOPERATIVE SCENARIOS

CHAPTER 3.   DOMINATING STRATEGIES

1.   UNDOMINATED STRATEGIES

In the n-player game, $G = (X_i, u_i, i = 1, \ldots, n)$, we study decentralized behavior: The players do not communicate. Outcomes are not biased for historical reasons (like records of previous plays, or some starting position).   On the contrary, all strategies are equally likely to occur so that discrimination among them must follow from endogenous arguments for rationality.   We also assume that a player is aware of his or her utility function but not necessarily of those of others.

Whenever the players cannot communicate at any time, they make strategic decisions independently; each player, unaware of the choices of other players and knowing that his own choice will not affect theirs, confines his choice to undominated strategies.

Definition 1.

In game $(X_i, u_i, i = 1, \ldots, n)$ say that strategy x_i of Player i *dominates* his strategy y_i if we have

$$u_i(y_i, x_{-i}) \leq u_i(x_i, x_{-i}) \quad \text{all } x_{-i} \in X_{-i}$$

$$u_i(y_i, x_{-i}) < u_i(x_i, x_{-i}) \quad \text{at least one } x_{-i} \in X_{-i}$$

We denote by $D_i(u_i; X)$ the set of Player i's *undominated* strategies:

$$x_i \in D_i(u_i, X) \iff \nexists y_i \in X_i : y_i \text{ dominates } x_i$$

Player i's strategy x_i dominates y_i if, no matter how the "rest of the world" $N\setminus\{i\}$ behaves, it never pays more to play y_i than x_i and for some feasible strategical choice of $N\setminus\{i\}$ it pays strictly more to play x_i than y_i. This suggests that Player i should pick his strategy within $D_i(u_i, X)$.

Notice that Player i's undominated strategies, just like his prudent strategies, depend on the various strategy sets and Player i's preferences only, not on other players' preferences. Like prudent strategies again, undominated strategies exist under familiar topological assumptions.

Lemma 1.

Let for all $i = 1, \ldots, n$, X_i be a compact space and u_i be a continuous function on X_N. Then $P_i(u_i; X)$, the set of Player i's prudent strategies, and $D_i(u_i; X)$, the set of his undominated strategies, are both nonempty and intersect:

$$P_i(u_i; X) \cap D_i(u_i; X) \neq \emptyset$$

Proof.

Step 1 D_i is nonempty

Assume that two strategies x_i, $y_i \in X_i$ are equivalent if $u_i(x_i, x_{-i}) = u_i(y_i, x_{-i})$ for all $x_{-i} \in X_{-i}$. Consider the binary relation R on strategy set X_i:

$x_i \, R y_i \; < \; = \; > \; \{x_i$ dominates $y_i\}$ or $\{x_i$ and y_i are equivalent$\}$

Clearly R is reflexive and transitive. By the topological assumptions, every chain (subset of X_i completely ordered by R) has a majoring element in X_i. Take x_i^μ to be such a chain, and pick x_i^* to be the limit of some converging subsequence $x_i^{\mu_n}$. Then $x_i^* \, R \, x_i^\mu$ holds for any μ. Namely, $u_i(x_i^\mu, x_{-i}) \le u_i(x_i^{\mu_n}, x_{-i})$ holds for all x_{-i} and any integer n such that μ_n is above μ in the chain. Taking the limit w.r.t. n yields the claim.

Thus we may apply Zorn's lemma. There exists a maximal element $x_i \in X_i$ with respect to R. Such a strategy is undominated.

Step 2 P_i is nonempty and compact

Since u_i is continuous on X, the real-valued function on X_i

$$\theta(y_i) = \inf_{x_{-i} \varepsilon X_{-i}} u_i(y_i, x_{-i})$$

is upper semicontinuous on X_i and therefore reaches its maximum over a nonempty compact set $P_i(u_i; X)$, namely, the set of Player i's prudent strategies.

Step 3 D_i and P_i intersect

Consider the game $H = (Y_j, u_j, j = 1, \ldots, n)$, where $Y_j = X_j$ for all $j \neq i$, and $Y_i = P_i(u_i; X)$.

By Step 1, Player i has at least one undominated strategy x_i in H. Suppose x_i is dominated by y_i in the original game (X_i, u_i). Thus, $\forall\ x_{-i} \varepsilon X_{-i}\ u_i(x_i, x_{-i}) \leq u_i(y_i, x_{-i}) \Rightarrow \theta(x_i) \leq \theta(y_i)$. This implies $\theta(y_i) = \sup_{z_i \varepsilon X_i} \theta(z_i)$ and $y_i \varepsilon P_i(u_i)$, contradicting our assumption that x_i is undominated in H. Thus x_i belongs to $P_i(u_i) \cap D_i(u_i)$. QED.

Notice that under the premises of Lemma 1, the set P_i of prudent strategies is compact, while D_i, the set of undominated strategies, might not be. Exercise 8 elaborates on this point.

2. DOMINATING STRATEGIES

Whenever a player happens to have a dominating strategy, the no-communication assumption makes it his unambiguously preferred choice.

Definition 2.

In game $G = (X_i, u_i, i = 1, \ldots, n)$ a strategy x_i of Player i is *dominating* if we have

for all $y_i \varepsilon X_i$, $x_{-i} \varepsilon X_{-i}$: $u_i(y_i, x_{-i}) \leq u_i(x_i, x_{-i})$

We denote by $D_i(u_i; X)$ the set of Player i's dominating strategies. An *equilibrium in dominating strategies* is an outcome x such that for all i, strategy x_i is dominating for Player i.

Example 1. First- and second-price auction.

An object is to be allocated among n agents. Its value is zero to the seller (who is not a player in the subsequent game) and a_i to Agent i. We order the agents so that

$$0 \leq a_n \leq a_{n-1} \leq \ldots \leq a_2 \leq a_1$$

In the first-price auction each agent independently bids— denote x_i Agent i's bid — next the highest bidder wins the object and pays his bid. Hence the game

$$X_1 = \ldots = X_n = [0, +\infty[$$

For any $x \in X_{\{1, \ldots, n\}}$ we set $w(x) = \{i/x_i = \sup_{1 \le j \le n} x_j\}$

$$u_i(x) = a_i - x_i \qquad \text{if } i = \inf_{j \in w(x)} j$$

$$= 0 \qquad \text{otherwise}$$

Note that ties are broken in favor of the player who values more the object, a convention with negligible effect. Observe that a_i, the sincere strategy of Agent i, dominates every strategy x_i such that $a_i \le x_i$: it never pays to overbid. Namely, $u_i(x_i, x_{-i}) \le 0 = u_i(a_i, x_{-i})$ for all x_{-i}. Thus $\mathcal{D}_i(u_i) = [0, a_i]$. However Player i has no dominating strategy unless $0 = a_i$. It might pay to underbid. This feature is analyzed in detail in the next chapter (see Exercise 2, Chapter 4).

In the <u>second-price</u> auction or Vickrey's auction, the highest bidder wins the object but is charged only the second highest price. Formally,

$$X_1 = \ldots = X_n = [0, +\infty[$$

For any $x \in X_{\{1, \ldots, n\}}$ we write $\tilde{x}_{-i} = \sup_{\substack{1 \le j \le n \\ j \ne i}} x_j$

$$u_i(x) = a_i - \tilde{x}_{-i} \qquad \text{if } i = \inf_{j \in w(x)} j$$

$$= 0 \qquad \text{otherwise}$$

We claim that <u>the sincere strategy a_i is a dominating</u> <u>strategy of Player i</u>. Fix an outcome $x \in X_N$ and distinguish two cases.

<u>Case 1.</u>: Player i wins the auction at outcome (a_i, x_{-i}). This implies $\tilde{x}_{-i} \le a_i$. Observe that $u_i(x)$ is $a_i - \tilde{x}_{-i}$ if i still wins the object at x and zero otherwise. Hence,

$u_i(x_i, x_{-i}) \le a_i - \tilde{x}_{-i} = u_i(a_i, x_{-i})$.

<u>Case 2</u>: Player i does not win the auction at (a_i, x_{-i}). Then $a_i \le \tilde{x}_{-i}$ and $u_i(x_i, x_{-i})$ is $(a_i - \tilde{x}_{-i})$ or zero. Therefore: $u_i(x_i, x_{-i}) \le 0 = u_i(a_i, x_{-i})$.

One can easily check that no other strategy of Player i is dominating. Thus $D_i(u_i) = \{a_i\}$. Every agent has a noncooperative incentive to report truthfully his or her valuation of the object. At the equilibrium in dominating strategies, Player 1 gets the object and pays a_2.

Several facts narrow down the scope for dominating strategy equilibria. The most serious is that they rarely exist. A dominating strategy is but a common solution to the programs $\max_{x_i} u_i(x_i, x_{-i})$ for all values of the parameter x_{-i}.

Second, they may yield Pareto-dominated outcomes:

outcome $x \in X_N$ Pareto dominates outcome $y \in X_N$ if $u_i(y) \le u_i(x)$, all $i = 1, \ldots, n$, <u>and</u> at least one inequality is strict.

In Vickrey's auction the dominating strategy equilibrium is Pareto dominated. If all agents bid $x_j = 0$, Agent 1 gets the object for free. Another famous instance of this situation is the prisoner's dilemma.

Example 2. <u>Prisoner's dilemma.</u>

Each of the two players is endowed with two strategies A, P where A stands for aggressive and P for peaceful. We assume that "peace" (both players are peaceful) is better for both players than "war" (both players are aggressive) but uni-lateral aggression (one player is aggressive while the other is peaceful) is profitable to the aggressor. A typical payoff structure is the following.

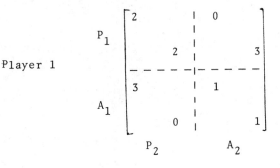

Player 2

Player 1's strategies are the rows of the matrix; Player 2's strategies are its columns. An entry of the 2 x 2 matrix is an outcome of the game. The northwest number is Player 1's utility and the southeast number is Player 2's utility. For instance, $u_1(A_1, P_2) = 3$.

For both players strategy A dominates strategy P. Hence (A_1, A_2) is the (unique) dominating strategy equilibrium. Therefore war is the postulated noncooperative outcome. However (P_1, P_2) (peace) provides a better utility level to both players. Therefore noncooperative selfish rationality conflicts with collective interest arguments. Collective interests demand sticking to the peaceful strategy; yet, if the players do not communicate, so that bilateral threats of the form "I shall be peaceful just as long as you are peaceful" cannot emerge, war is the likely outcome. Decentralization of the strategic choices has a high collective cost.

All noncooperative equilibria share the defect illustrated in the prisonner's dilemma: Non-cooperative scenarios ignore common-interest arguments.

The third poor feature of dominating strategy equilibria is the possible multiplicity of their utility vectors. If a player has several dominating strategies, they are equivalent to him, perhaps not to the others. In our example a player's strategy influences only his partner's utility.

Example 3. Do not forget the favor.

		Nice to Player 1	Not nice to Player 1
Player 1	Nice to Player 2	1 1	0 1
	Not nice to Player 2	1 0	0 0

Player 2

Every outcome is a dominating strategy equilibrium, yet only one outcome is Pareto optimal.

3. EXERCISES

a) Examples of games with dominating strategy equilibrium

1) A quantity setting oligopoly (Case [1979])

Suppose that the price of a typical satiable good, say mineral water, goes to ce^{-S} where S is the total supply. If n costless producers control the quantities x_1, ..., x_n of mineral water that they supply, we obtain the following game

$$u_i(x_1, \ldots, x_n) = cx_i e^{-(x_1 + \ldots + x_n)}$$

Compute the equilibrium in dominating strategies and comment upon the assumed form of the inverse demand.

2) A price setting duopoly (Case [1979])

Two duopolists offer two substitutable goods. If they set the prices p_1, p_2 the corresponding demands are

$$d_1 = \left[\frac{p_2}{p_1}\right]^{\alpha_1} \text{ units of the good produced by Player 1}$$

$$d_2 = \left[\frac{p_1}{p_2}\right]^{\alpha_2} \text{ units of the good produced by Player 2}$$

We assume $\alpha_i \geq 1$, $i = 1$, 2. Suppose, in addition a constant return to scale technology for both producers. Thus the following normal form game.

$$u_i(p_1, p_2) = (p_i - c_i) \cdot d_i$$

where c_i is constant.

Compute the equilibrium in dominating strategies of this game and comment on the assumed form of the demand functions.

3) A Colonel Blotto game

There are ten locations with respective value $a_1 < \ldots < a_{10}$. Player i ($i = 1$, 2) is endowed with n_i soldiers ($n_i < 10$) and must allocate them among the locations. To each particular location he can allocate no more than one soldier. The payoff at location p is a_p to the player whose soldier is unchallenged, and $-a_p$ to his opponent, unless both have a soldier at p or no one has, in which case the payoff is zero to both. The total payoff is obtained by summing up local payoffs.

Show that this game has a unique dominating strategy equilibrium.   What if some of the a_p coincide?

4)   Games with linear payoffs (Moulin [1979])

   a)   Consider first a two-plaer game with strategy sets $X_1 = X_2 = [-1, +1]$ (the real interval) and linear utility functions

$$u_1(x_1, x_2) = ax_1 + bx_2, \quad u_2(x_1, x_2) = cx_1 + dx_2$$

where a, b, c, d are four fixed real numbers. Clearly each player has a dominating strategy (unique if a — resp d — is nonzero).

   We are interested in configurations where a prisoner's dilemma arises, namely, the dominating strategy equilibrium is Pareto dominated. Prove that this is the case iff

$$a.c < 0, \quad b.d < 0, \quad \text{and} \quad 1 < \frac{b.c}{a.d}$$

Hint: Draw a picture of the feasible utility set, namely $(u_1, u_2)(X_1 \times X_2)$.

   b)   Now we have a n-person game with $X_i = [-1, +1]$, all $i = 1, \ldots, n$, and Player i's utility function

$$u_i(x) = \sum_{j=1}^{n} a^j_i x_j$$

We assume $a^i_i \neq 0$, all $i = 1, \ldots, n$. Denote by x^* the unique dominating strategy equilibrium. Prove the equivalence of the three following statements:

i) From the prisonner's dilemma effect, x^* is a Pareto-dominated outcome.

ii) There is an outcome $y \in X_N$ such that

$$\{a^i_i y_i > 0 \quad \text{and} \quad u_i(y) < 0\} \quad \text{all } i = 1, \ldots, n$$

This says that each player would be better off by using y_i alone (i.e., while others stay put at strategy zero), but all will be worse off when they all use y_i.

iii) For all $p, q \in R^n_+$ (vectors with nonnegative components); the system

$$\{\sum_{i=1}^{n} p_i a^j_i = q_j a^j_j\} \quad \text{all } j = 1, \ldots, n$$

implies $p = q = 0$.

As an example, consider the game "provision of a public good":

$$u_i = -\alpha_i x_i + \lambda \{\sum_{j=1}^{n} x_j\} \quad \alpha_i > 0$$

Interpret x_i as the (positive or negative) effort spent by i for the common welfare, and $-\alpha_i$ as his marginal disutility for the effort. For which values of λ do we have a prisonner's dilemma effect?

b) <u>Abstract games</u>

5) Give an example of a two-person game with a Pareto-optimal-dominating strategy equilibrium, but which is <u>not</u> inessential.

6) Give an example of a 3 x 3 two-person game (i.e., each player has exactly three strategies) where no strategy is dominated and no two are equivalent for any player ($\mathcal{D}_i = X_i$, i = 1, 2) yet the game is inessential. Can you find a similar example with a 2 x 3 or a 2 x 2 game?

7) Recall that strategies x_i and y_i of Player i are <u>equivalent</u> if they are not distinguishable in Player i's opinion:

$$\forall\ x_{-i}\ \varepsilon\ X_{-i}: \quad u_i(x_i,\ x_{-i}) = u_i(y_i,\ x_{-i})$$

Suppose that game $(X_j,\ u_j;\ j = 1,\ \ldots,\ n)$ is finite or X_j are compact and u_j continuous. Then the three following properties are equivalent. i) A dominating strategy for Player i exists: $D_i(u_i) \neq \emptyset$, ii) All strategies in $\mathcal{D}_i(u_i)$ are equivalent, iii) $D_i(u_i) = \mathcal{D}_i(u_i)$. Deduce that a player has a unique undominated strategy if and only if he has a unique dominating strategy.

8) <u>Topological properties of D_i and $\mathcal{D}_i$</u>

For all i = 1, ..., n, let X_i be a compact set and u_i be a continuous function on X. Show that the sets $\mathcal{D}_i(u_i)$ of undominated strategies are not necessarily closed. What about the sets $D_i(u_i)$ of dominating strategies? <u>Hint</u>: Consider the following game.

$$X_1 = X_2 = [0, 1] \quad u_1(x_1, x_2) = 0 \quad \text{if } \frac{x_1}{2} + x_2 \geq 1$$

$$u_1(x_1, x_2) = x_1 - \frac{x_1 x_2}{1 - \frac{x_1}{2}} \qquad \qquad \text{if } \frac{x_1}{2} + x_2 \leq 1$$

u_2 is arbitrary.

Prove that u_1 is continuous and yet $\mathcal{D}_1(u_1)$ is not closed.

9) In Borel's poker model (Example 2, Chapter 1) justify the restriction to <u>simple</u> strategies. Take Y_i to be the borelian σ algebra of $[0, 1]$ and interpret strategy $y_i \in Y_i$ as Player i will bid iff $m_i \in y_i$. Prove that any strategy in $Y_i \setminus X_i$ is dominated by a strategy in X_i.

CHAPTER 4. SOPHISTICATED AND PERFECT EQUILIBRIUM

1. SUCCESSIVE ELIMINATION OF DOMINATED STRATEGIES

A dominating strategy is the most advantageous non-cooperative strategy one can employ no matter how much that player knows of the other players' utilities. By contrast, the scenarios proposed in this chapter rely on the complete information assumption. That is, each player is aware of the entire normal form of the game, including other player's preferences. Adhering to this assumption, noncooperating players mutually anticipate each other's strategic choices. Each player expects all other players to eliminate their dominated strategies. This, in turn, may generate new dominated strategies, and so on.

Example 1. Guess the average.

Each player picks an integer between 1 and 999. Let

$$\bar{x} = \frac{1}{n} \sum_{i=1}^{n} x_i$$

be the average answer. The winners are those players whose ballot is closest to $2/3 \ \bar{x}$.

Every strategy above 666 is dominated (by 666). Thus if all players use undominated strategies only, each strategy set shrinks from $\{1, \ldots, 999\}$ to $\{1, \ldots, 666\}$. The argument can be repeated. Every strategy above 444 is now dominated (by 444), so nobody will use any such strategy, and so on. Finitely many repetitions of the argument force the conclusion that each player chooses $x_i = 1$.

Definition 1 (Moulin [1979]).

In the n-player game $G = (X_i, u_i, i = 1, \ldots, n)$, the successive elimination of dominated strategies is made up of the sequences

$$X_i = X_i^0 \supset X_i^1 \supset \ldots \supset X_i^t \supset X_i^{t+1} \ldots \text{ all } i = 1, \ldots n$$

where $X_i^{t+1} = D_i(u_i; X^t)$.

We denote $X_i^\infty = \bigcap_{t=0}^{\infty} X_i^t$. We shall say that G is dominance solvable if X_N^∞ is nonempty and Player i's utility u_i does not depend on x_i on X_N^∞:

$$u_i(x_i, x_{-i}) = u_i(y_i, x_{-i}) \text{ all } x_i, y_i \in X_i^\infty,$$
$$x_{-i} \in X_{-i}^\infty$$

In this case we call X_N^∞ *the set of* <u>*sophisticated*</u>
<u>*equilibria*</u> *of* G.

Typically if X_i^∞ is a singleton for all i = 1, ..., n,
then G is dominance solvable. The behavior of sophisticated
players is both static and decentralized. The game is played
once after each player has performed, independently, the
successive elimination of dominated strategies; when this
algorithm converges (i.e., the game is dominance solvable)
our players eventually pick an unambiguous equilibrium
strategy. In game $(X_i, u_i, i = 1, ..., n)$ all Player i's
strategies are equivalent to him. True, this does not imply
that they are equivalent to other players (see Example 3,
Chapter 3) yet, in general (e.g., when utility functions are
one to one on X_N), we can expect the sophisticated equilibrium
to be a singleton. In dominance solvable games, sophisticated
behavior is essentially deterministic.

Our next example displays implicit cooperation at
sophisticated equilibrium.

Example 2. <u>Plurality voting with ties</u> (Farqharson [1969]).
Among three candidates {a, b, c}, a society 1, 2, 3 must
elect one. The voting rule is plurality voting and Player 1
breaks ties. In other words, the strategy sets (or message
set) are $X_1 = X_2 = X_3 = \{a, b, c\}$, and, if the agents cast
the votes (x_1, x_2, x_3), the elected candidate is

$$\pi(x_1, x_2, x_3) = x_2 \qquad \text{if } x_2 = x_3$$

$$= x_1 \qquad \text{if } x_2 \neq x_3$$

Suppose now that the utility of players for the various candidates display a Condorcet effect,

$$u_1(c) < u_1(b) < u_1(a)$$

$$u_2(b) < u_2(a) < u_2(c)$$

$$u_3(a) < u_3(c) < u_3(b)$$

Player 1 has a dominating strategy (vote for a) hence his ballot is predictable at once. Player 2 and 3, however, have only one dominated strategy, namely, to vote for their least-preferred candidate. Thus, after one round of elimination, we have

$$X^1_1 = \{a\}, \; X^1_2 = \{a, c\}, \; X^1_3 = \{b, c\}$$

In game $(X^1_i, u_i\pi, i = 1, 2, 3)$ we note that outcome b cannot be elected any longer, hence Player 3's strategy b is now dominated by his strategy c. By the same token, Player 2's strategy a is now dominated by his strategy c. Thus, after two rounds of elimination, outcome (a, c, c) emerges as the sophisticated equilibrium. Player 1's privilege as the tie breaker yields eventually his worst outcome!

How realistic is the strategic behavior implied by the concept of sophisticated equilibrium? It presupposes that each and every Player either computes all dominated strategies of all players or performs the successive elimination as long as necessary, on the assumption that every other player is equally patient. The first point does not cause much trouble. In fact, when the game is dominance solvable, nothing would be changed if the elimination process was to take place successively, in arbitrary order; the players could also drop only part of the dominated strategies. These robustness properties, however, are not easy to prove (see Lemma 1 below).

The second point raises a more serious objection. As Example 1 above makes clear, the successive elimination can be arbitrarily long. In practice, most players do not perform the elimination forever out of fear that other players are not rational enough to do so. This is what experimental evidence of the "guess the average" game suggests. That is, if 20 players are involved, you do not expect that all of them will perceive the geometrical shrinking of strategy sets. In fact, they do not, and the winning guess usually lies between 100 and 200.

Our last example shows that elimination of dominated strategies can yield a deterministic noncooperative outcome even when the game is not dominance solvable.

Example 3.   The Steinhaus method to share a cake.

Let $[0, 1]$ be a nonhomogeneous cake to be divided among two players. The utility of Player 1 for a share $A \subset [0, 1]$ is worth

$$v_1(A) = \int_A (\frac{3}{2} - x)\,dx$$

The utility of Player 2 for a share $B \subset [0, 1]$ is worth

$$v_2(B) = \int_B (\frac{1}{2} + x)\,dx$$

When time runs from $t = 0$ to $t = 1$, a knife is moved at speed 1 from $x = 0$ to $x = 1$. Both players can stop it at any time. If the knife is stopped at time t by Player i, this player gets the share $[0, t]$ whereas the other player gets $[t, 1]$. Thus the strategy sets are $X_1 = X_2 = [0, 1]$ where strategy x_i means that Player i will stop at time $t = x_i$ unless the other player did so before that time. The utility functions are given by

$$u_1(x_1, x_2) = v_1([0, x_1]) = \frac{3}{2} x_1 - \frac{x_1^2}{2} \quad \text{if } x_1 \leq x_2$$

$$= v_1([x_2, 1]) = 1 - \frac{3}{2} x_2 + \frac{x_2^2}{2} \quad \text{if } x_2 < x_1$$

$$u_2(x_1, x_2) = v_2([x_1, 1]) = 1 - \frac{1}{2} x_1 - \frac{1}{2} x_1^2 \quad \text{if } x_1 \leq x_2$$

$$= v_2([0, x_2]) = \frac{1}{2} x_2 + \frac{1}{2} x_2^2 \quad \text{if } x_2 < x_1$$

The strategy t_i, $i = 1, 2$ defined by

$$v_i([0, t_i]) = v_i([t_i, 1]) = \frac{1}{2}$$

is the unique prudent strategy of Player i. As a result, we have

$$\phi_i(x_i) = \inf_{x_j \in [0,1]} u_i(x_1, x_2) = \inf \{v_i([0, x_i]), v_i([x_i, 1])\}$$

so ϕ_i reaches its maximum when both shares $[0, t_i]$ and $[t_i, 1]$ are equivalent to Player i. By our specific choice of v_i we have clearly

$$\frac{3 - \sqrt{5}}{2} = t_1 < t_2 = \frac{\sqrt{5} - 1}{2}$$

Prudent behavior implies that Player 1 gets the share $[0, t_1]$. Hence his secure utility level $\alpha_1 = v_1([0, t_1])$, whereas Player 2 is more lucky: $v_2([t_1, 1]) > v_2([t_2, 1]) = \alpha_2$. In fact, the prudent outcome allocates the whole surplus available from (α_1, α_2) to Player 2:

$$v_2([t_1, 1]) = \max \{u_2(x_1, x_2)/u_1(x_1, x_2) \geq \alpha_1\}$$

The first round of elimination of dominated strategies gives

$$D_i(u_i) = [t_i, 1]$$

Indeed a strategy x_i that would stop the knife before t_i is dominated by the prudent strategy t_i (as an exercise, check this claim). With complete information, Player 1 can compute t_2. Given that Player 2 will not stop the knife before t_2, a second round of elimination of dominated strategies allows him to drop the strategies in $[t_1, t_2]$. Indeed, for all $x_1 \in [t_1, t_2]$ we have

$$v_1([0, x_1]) = u_1(x_1, x_2) < u_1(t_2, x_2) = v_1([0, t_2])$$

$$\text{all } x_2 \in [t_2, 1]$$

Thus $X_1^2 = [t_2, 1]$, and no further elimination of dominated strategies is possible for either player:

$$X_1^t = X_2^t = [t_2, 1] \qquad \text{all } t \geq 2$$

Since no two strategies are equivalent to Player i in $X_1^t \times X_2^t$, we conclude that our game is <u>not</u> dominance solvable. However, the restricted game $([t_2, 1] , [t_2, 1], u_1, u_2)$ is <u>inessential</u>. Both players have the same prudent strategy $x_1 = x_2 = t_2$:

$$\tilde{\alpha}_1 = \inf_{x_2 \epsilon [t_2, 1]} u_1(t_2, x_2) = v_1([0, t_2]) > \inf_{x_2 \epsilon [t_2, 1]} u_1(x_1, x_2)$$

$$= v_1([x_1, 1]), \qquad \text{all } x_1 > t_2$$

$$\tilde{\alpha}_2 = \inf_{x_1 \epsilon [t_2, 1]} u_2(x_1, t_2) = v_2([t_2, 1]) > \inf_{x_1 \epsilon [t_2, 1]} u_2(x_1, x_2)$$

$$= v_2([x_2, 1]) \qquad \text{all } x_2 > t_2$$

Furthermore, $(\tilde{\alpha}_1, \tilde{\alpha}_2)$ is a Pareto-optimal utility vector, as is clear from Figure 1. We conclude from Theorem 1, Chapter 2 that $x_1 = x_2 = t_2$ is an optimal strategy for both players in the 2-reduced game, which makes (t_2, t_2) the predicted outcome when Player 1 has complete information on Player 2's utility. Notice that $(\tilde{\alpha}_1, \tilde{\alpha}_2) = (v_1([0, t_2]), v_2([t_2, 1])$ allocates the whole surplus to Player 1:

$$v_1([0, t_2]) = \max \{u_1(x_1, x_2)/u_2(x_1, x_2) \geq \alpha_2\}$$

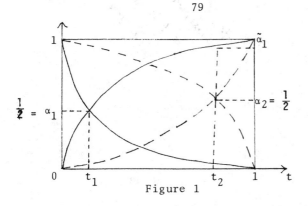

Figure 1

The two curves crossing at (t_1, α_1) are the graphs of $v_1([0, t])$ and $v_1([t, 1])$. The two dotted curves crossing at (t_2, α_2) are the graphs of $v_2([0, t])$ and $v_2([t, 1])$.

2. GAMES IN EXTENSIVE FORM AND KUHN'S ALGORITHM

For normal form games, no sufficient conditions for dominance solvability are known (an exception is Theorem 3, Chapter 6). For games in extensive form, on the contrary, Kuhn's theorem (Theorem 3, Chapter 1) is generalized into a dominance solvability statement.

The following definition generalizes that of Chapter 1, Section 3. The same notations are used.

Definition 2. Games in extensive form.

An n-player game in extensive form is defined by a finite tree $\Gamma = (M, \sigma)$, a partition $(M_i, i = 1, \ldots, n)$ of the set $M \backslash T(M)$ of nonterminal nodes of Γ, for each Player $i = 1, \ldots, n$, a utility function defined on

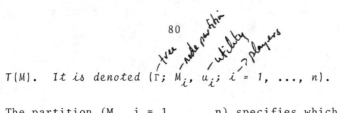

$T(M)$. *It is denoted* $(\Gamma; M_i, u_i; i = 1, \ldots, n)$.

The partition $(M_i, i = 1, \ldots, n)$ specifies which player has the move at each nonterminal node. We assume that at each node, the player who moves <u>knows</u> the current node as well as the entire definition of the game (including other players' utilities). This understanding is usually called the <u>perfect-information assumption</u> in extensive games. Here is an example with three players.

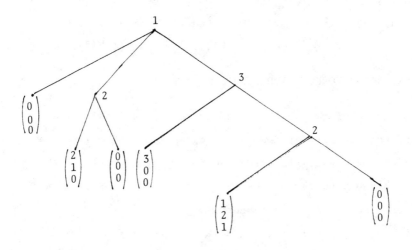

Figure 2

in each bracket Player 1's payoff is on top, 2's is middle, and 3's is bottom.

A game in extensive form has a canonical normal-form representation $(X_i, \tilde{u}_i, i = 1, \ldots, n)$ where a strategy $x_i \in X_i$ specifies Player i's move at each node in M_i. Thus X_i is made up of the mappings x_i from M_i into M, such that

$$x_i(m) \in \sigma^{-1}(m) \qquad \text{all } m \in M_i$$

To each n-tuple $x = (x_1, \ldots, x_n)$ is associated a unique play, i.e., a path from the origin m_o to some terminal node $m(x)$. We set $\tilde{u}_i(x) = u_i(m(x))$, all $i = 1, \ldots, n$.

For instance, in the game of Figure 2, Player 1 has three strategies, Player 2 has four, and Player 3 has two.

Kuhn's theorem states that the normal-form games derived from games in extensive form are, in general, dominance solvable, and their sophisticated equilibrium outcome is computed by a simple backward algorithm. To clarify this result we analyze the game in Figure 2.

In both nodes where he has the move, Player 2 makes an ultimate choice. In both cases he prefers to move left. Anticipating this choice, Player 3 faces now the reduced tree

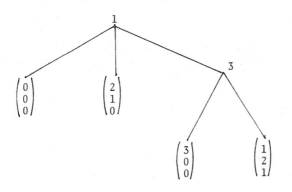

whereupon he would move right if Player 1 ever let him play. But Player 1 foresees this and decides to play middle. The equilibrium play is as follows. Player 1 moves to the middle. Next Player 2 moves left.

To reinforce intuition for Kuhn's backward algorithm we offer a somewhat more elaborate example.

Example 4. Voting by veto.

Let A = {a, b, c, d} be a set of four candidates from which the voters {1, 2, 3} must elect one. The following voting rule is in order. Starting with Player 1, each player successively vetoes one of the nonvetoed candidates. The remaining candidate is elected. Individual preferences are as follows.

$$u_1(d) < u_1(c) < u_1(b) < u_1(a)$$

$$u_2(c) < u_2(d) < u_2(a) < u_2(b) \tag{1}$$

$$u_3(c) < u_3(a) < u_3(b) < u_3(d)$$

The tree in Figure 3 describes the extensive-form game. It is understood that the corresponding utilities given by (1) should be attached to a terminal node. Observe first that Player 3 has a dominant strategy. Indeed, his choice among the last two nonvetoed candidates is final. Hence, he should always veto sincerely. This leads to the reduced

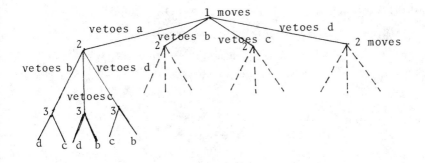

Figure 3

game shown in Figure 4.

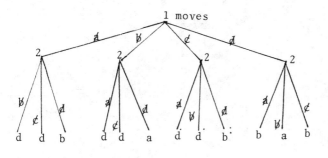

Figure 4

Player 1 has only one prudent strategy, namely, veto d since <u>only</u> this move <u>guarantees</u> the defeat of d — no matter how Player 2 and 3 behave. However a sophisticated Player 1, aware of Player 2 and 3's preferences, realizes that vetoing d yields the inevitable election of b, for Player 2 and 3 share the same opinion of a, b, c. Player 1 does better by vetoing b, after which Player 2 unambiguously prefers to veto d — resulting in the election of a — than to veto a or c which would result in the election of d.

The backward algorithm generalizes easily to a large class of extensive-form games. We say that the game $(\Gamma, M_i, u_i, i = 1, \ldots, n)$ satisfies the <u>one-to-one assmumption</u> if for any terminal nodes m, m' $\in$ T(M) and any Player i

$$\{u_i(m) = u_i(m')\} \Rightarrow \{u_j(m) = u_j(m') \text{ all } j = 1, \ldots, n\} \quad (2$$

Definition 3. Kuhn's reduction algorithm.

Let L(M) be the subset of those nodes m of which all successors are terminal nodes. Thus,

$$m \in L(M) \iff \emptyset \neq \sigma^{-1}(m) \subset T(M)$$

Then the reduced game of G is $G^* = (\Gamma^*, M_i^*; u_i^*, i = 1, \ldots, n$ where: set of nodes $M^* = M \backslash T_0(M)$ where $T_0(M) = \{m \in T(M)/\sigma(m) \in L(M)\}$, mapping $\sigma^* =$ restriction of σ to M^*; terminal nodes of (M^*, σ^*): $T(M^*) = L(M) \cup \{T(M) \backslash T_0(M)\}$, $M_i^* = M_i \cap \{M^* \backslash T(M^*)\}$, hence $(M_i^*)_{i \in N}$ partition $M^* \backslash T(M^*)$; utility u_i^*: if $m \in T(M) \backslash T_0(M)$: $u_i^*(m) = u_i(m)$; if $m \in L(M)$ and $m \in M_j$, then $u_i^*(m) = u_i(m_j)$ where m_j is an optimal successor to m for Player j: $u_j(m_j) = \sup_{m' \in \sigma^{-1}(m)} u_j(m')$.

We claim that if Γ is a tree with length ℓ, then Γ^* has length $(\ell - 1)$ (this claim is the subject of Step 1 in the subsequent proof). Kuhn's algorithm consists of ℓ successive reductions of game G. After these ℓ reductions the game tree $(M^{*\ell}, \sigma^{*\ell})$ has zero length, namely, $M^{*\ell} = \{m_0\}$, and $\sigma^{*\ell}$ is the identity mapping. We denote by $\beta_i = u_i^{*\ell}(m_0)$ the unique payoff of Player i in $G^{*\ell}$.

Theorem 1.

Let $G = (\Gamma, M_i, u_i, i = 1, \ldots, n)$ be an n-player game in extensive form satisfying (2). Then the normal form of G is dominance solvable, and its sophisticated equilibrium payoffs $(\beta_1, \ldots, \beta_n)$ are computed by Kuhn's algorithm.

Proof.

Step 1   The length of Γ^* is that of Γ minus one
We pick a node m such that:

$$\sigma^\ell(m) = m_0, \quad \sigma^{\ell-1}(m) \neq m_0 \tag{3}$$

Such a node exists by definition of ℓ. Suppose $\sigma(m) \notin L(M)$. Then there exists m' ϵ M\T(M) such that $\sigma(m') = \sigma(m)$. Picking any m'' ϵ $\sigma^{-1}(m')$, we get $\sigma^2(m'') = \sigma(m)$. Hence, $\sigma^\ell(m'') = \sigma^{\ell-1}(m) \neq m_0$, which is a contradiction. We have proved $\sigma(m)$ ϵ L(M). Since m is clearly a terminal node, we conclude m ϵ $T_0(M)$. Thus no node such that (3) holds can be in M*, and therefore (M*, σ*) has length at most $(\ell - 1)$. It is

just as easy to prove that (M*, α*) has <u>exactly</u> length
ℓ - 1.

Step 2

Let $(X_i, \tilde{u}_i, i = 1, \ldots, n)$ be the normal form game G. We
have to prove that the game $(X_i, \tilde{u}_i, i \in N)$ is dominance
solvable and yields the same equilibrium payoff as Kuhn's
algorithm. The reduction of G into G* (Definition 3) amounts
to dropping all nodes of $T_o(M)$, thus making the nodes of L(M)
terminal, and assigning to a node m in $L(M) \cap M_i$ the payoff
vector that would result if Agent i were to take the optimal
move at m. Henceforth, if A_i denotes the following subset
of X_i,

$$A_i = \{x_i \in X_i / \forall m \in L(M) \cap M_i, u_i(x_i(m)) = \sup_{m' \in \sigma^1(m)} u_i(m')\}$$

Then the N-normal-form game $(A_i, \tilde{u}_i, i \in N)$ is isomorphic to
the reduced game G*. Notice that if i is indifferent between
two successors of $L(M) \cap M_i$, then by (2) all players are
indifferent as well, so that the corresponding strategies in
A_i can be identified. Observe that a strategy $x_i \in X_i \backslash A_i$ is
a dominated strategy for player i (indeed, x_i is dominated by
a strategy $y_i \in A_i$ that coincides with x_i on $M_i \backslash L(M)$). Thus
we have

$$\mathcal{D}_i(u_i) \subset A_i \subset X_i \tag{4}$$

We invoke now an auxiliary result.

Lemma 1 (Rochet [1980]).

Let $G = (X_i, u_i, i = 1, ..., n)$ be a _finite N-normal-form_ game such that _for any two outcomes_ x, x' ε X_N _we have_

$$[\exists \ i \ \epsilon \ N \ u_i(x) = u_i(x')] \Rightarrow \forall \ j \ \epsilon \ N \ u_j(x) = u_j(x') \qquad (5)$$

Next, _for all_ i ε N, _let_ A_i _be a subset of_ X_i _such that_ (4) _holds._

Then G _is dominance solvable iff_ $(A_i, u_i, i = 1, ..., n)$ _is and their sophisticated equilibrium payoffs coincide._

Applying Lemma 1 inductively yields

G d solvable <=> G* d solvable <=> G** d solvable

<=> ...

and these games all have the same sophisticated equilibrium payoffs. Since $G*^{\ell}$ is obviously d solvable with equilibrium payoffs $(\beta_i)_{i \in N}$, the proof of Theorem 1 is complete. QED.

Lemma 1 is a robustness result. During the successive elimination of dominated strategies, if some players are "lazy" and do not eliminate all their dominated strategies, or if the elimination is made sequentially (Player 1 first eliminates his dominated strategies, next, Player 2 does so, and so on), then the dominance solvability is preserved, and the sophisticated equilibrium payoffs are unaffected.

As a corollary to Theorem 1, we deduce Theorem 3, Chapter 1: Every finite two-person, zero-sum game in extensive form has a value. Indeed, a two-person, zero-sum game (in extensive form) satisfies assumption (2). Therefore it is dominance solvable by Kuhn's theorem. The desired conclusion follows from the following useful result.

Lemma 2.

Let $G = \{X_1, X_2; u_1\}$ be a two-person, zero-sum game in normal form with **finite** strategy sets. Then, a sophisticated equilibrium of G is a saddle point of u_1. Hence, a dominance solvable, two-person, zero-sum game has a value.

Proof.

Given two subsets $Y_1 \subset X_1$, $Y_2 \subset X_2$, we denote by $S(Y_1, Y_2)$ the (possibly empty) set of saddle points of the game (Y_1, Y_2, u_1). Next, we set $Z_i = \mathcal{D}_i(u_i; Y_1, Y_2)$. Let x_1, x_2 be a saddle point of game (Z_1, Z_2, u_1) and suppose that (x_1, x_2) is <u>not</u> a saddle point of game (Y_1, Y_2, u_1). For instance, there exists $y_1 \in Y_1$ such that

$$u_1(x_1, x_2) < u_1(y_1, x_2) \tag{6}$$

We set

$$Y_1(y_1) = \{y'_1 \in Y_1 / \forall\ y_2 \in Y_2\ u_1(y_1, y_2) \leq u_1(y'_1, y_2)\}$$

We claim that $Y_1(y_1)$ has a nonempty intersection with Z_1. For instance, if z_1 reaches the maximum of ϕ_1 over $Y_1(y_1)$ where

$$\phi_1(y'_1) = \sum_{y_2 \epsilon Y_2} u_1(y'_1, y_2)$$

then one checks that z_1 cannot be dominated in (Y_1, Y_2, u_1).
Therefore, z_1 is in $Y_1(y_1) \cap Z_1$ and from (6) we derive

$$u_1(x_1, x_2) < u_1(z_1, x_2)$$

This is a contradiction of our assumption that (x_1, x_2) is
a saddle point of (Z_1, Z_2, u_1).
We have just proved

$$S(Z_1, Z_2) \subset S(Y_1, Y_2)$$

Applying the inclusion iteratively yields

$$S(X_1, X_2) \supset S(X_1^1, X_2^1) \supset \dots \supset S(X_1^t, X_2^t)$$

Since X_i, $i = 1, \dots, n$ are finite, there is an integer t
such that $X_i^t = X_i^\infty$, $i = 1, \dots, n$. Since G is dominance
solvable, definition 1 then implies:

$$S(X_1^t, X_2^t) = X_1^t \times X_2^t = \text{sophisticated equilibria of G}$$

QED.

3. SUBGAME PERFECT EQUILIBRIUM

In this section we use the concept of Nash equilibrium,
which is the subject of the next chapter. We repeat the
necessary definition.

Consider a game in extensive form G = (Γ, M_i, u_i, i = 1, ..., n) satisfying the one-to-one assumption (2). A play is a path from the origin node m_o to some terminal node in T(M). At any node m along this path we could pause and think of the remaining strategic choices from the current node taken as the origin.

Definition 4.

For any nonterminal node m ε M\T(M), the subgame starting at m is G(m) = (Γ(m), M_i(m), u_i, i = 1, ..., n) where Γ(m) is the subtree of Γ with origin node m, and M_i(m) are the restrictions of M_i to the nodes of Γ(m).

An important feature of sophisticated equilibrium outcome of G is that its restriction to any subgame is still a sophisticated equilibrium.

Notations: The canonical normal form of a subgame G(m) is denoted (X_i(m), $\tilde{u}_i$, i = 1, ..., n). If x_i ε X_i is a strategy of Player i in G, we denote by x_i(m) the restriction of x_i to G(m). Finally we say that a strategy n-tuple y* in the normal form game (Y_i, $\tilde{u}_i$, i = 1, ..., n) is a <u>Nash</u> <u>equilibrium</u> if for all i = 1, ..., n,

for all y_i ε Y_i: $\tilde{u}_i(y_i, y^*_{-i}) \leq \tilde{u}_i(y^*)$

This definition is commented upon in Chapter 5.

Theorem 2 (*Selten* [1975]).

Let $G = (\Gamma, M_i, u_i, i = 1, \ldots, n)$ *be a game in extensive form satisfying the one-to-one assumption* (2), *with canonical normal form* $(X_i, \tilde{u}_i, i = 1, \ldots, n)$. *i) If x^* is a sophisticated equilibrium of G, then for any nonterminal node $m \in M \backslash T(M)$, its restriction $x^*(m)$ is a sophisticated equilibrium of the restricted game $G(m)$. ii) If x^* is a Nash equilibrium of G such that for any nonterminal node $m \in M \backslash T(M)$, the restriction $x^*(m)$ is a Nash equilibrium of $G(m)$, then x^* is a sophisticated equilibrium of G.*

Proof.

The natural projection p of X_i onto $X_i(m)$ $(x_i \to x_i(m))$ respects elimination of dominated strategies: $p(X_i^t) = [X_i(m)]^t$. The proof is left as an exercise. This in turn establishes Statement i.

Now we prove Statement ii. Take an outcome x^* satisfying the premises of ii) and use the notations of Step 2 in the proof of Theorem 1. Pick a node m in $L(M) \cap M_i$. The subgame starting at m is a one-player game involving Player i. Since $x^*(m)$ is a Nash equilibrium of it, we must have

$$u_i(x_i^*(m)) = \sup_{m' \in \sigma^{-1}(m)} u_i(m')$$

Thus $x_i^* \in A_i$, all $i = 1, \ldots, n$, and x^* can be identified with a strategy in G^*. Since the length of G^* is strictly less than that of G, we can prove statement ii by an obvious induction argument, left to the reader. QED.

Theorem 2 suggests the following definition.

Definition 5.

Let $G = (\Gamma, M_i, u_i, i = 1, \ldots, n)$ be a game in extensive form with the associated normal form $(X_i, \tilde{u}_i, i = 1, \ldots, n)$. Say that outcome $x^* \in X_N$ is a subgame perfect equilibrium of G if for any nonterminal node $m \in M \backslash T(M)$, the restriction $x^*(m)$ is a Nash equilibrium of the restricted game $G(m)$.

Theorem 2 states that whenever G satisfies the one-to-one assumption (2), subgame perfect equilibrium and sophisticated equilibrium are essentially the same notion. However if assumption (2) does not hold, this equivalence disappears. In fact, <u>all finite games in extensive form have at least one subgame perfect equilibrium,</u> whereas they may have no sophisticated equilibrium whenever assumption (2) fails to be true. The proof of the first claim is the subject of Exercise 9 below. The second claim is demonstrated by means of a variant of the game in Figure 2.

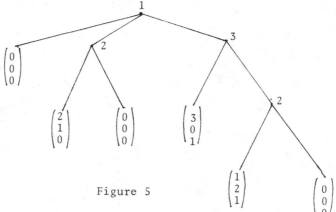

Figure 5

After anticipating Player 2's unambiguous choices we get the game shown in the accompanying figure. Player 3 is indifferent

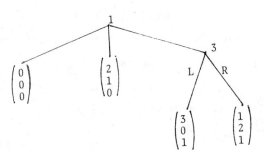

between moving left or right. Yet Player 1 is not indifferent to the fact that Player 3 has these two possible choices. Should Player 3 move left, Player 1 would move right. Should Player 3 move right, Player 1 would decide to move to the middle. As Player 1 is not indifferent to his two undominated strategies (moves to the middle or right), the game is not dominance solvable. However, two different subgame perfect equilibria are, respectively, x_1 = middle, x_2 = right with payoffs (2, 1, 0) and x_1 = right, x_3 = left, with payoffs (3, 0, 1) (Player 2 uses his dominating strategy in both cases). This example shows that subgame perfect equilibria do not yield a unique final utility vector when assumption (2) fails. Our next example helps confirm this point.

Example 5. A seemingly democratic sharing method.

Ten pirates (ranked 10 to 1 from the oldest to Benjamin, the youngest) share 100 gold coins. The oldest first submits an allocation to vote. If at least half of the pirates (including the petitioner) approves of this allocation, it is enforced. Otherwise, Pirate 10 is thrown out of the game. As a result, the nine-player version of the game will be played starting with a request by Pirate 9. For approval of this allocation the vote of at least four pirates among 8, ..., 1 is needed. And so on.

To work out Kuhn's algorithm, we need to know how a player reacts to request p which would give him the same number of coins as the proposal p' that he expects to be offered in response, should p be rejected. This amounts to select one subgame perfect equilibrium.

If only Pirates 2 and 1 are left, Pirate 2 keeps 100 coins for he does not need Pirate 1's approval. Think next of Pirate 3 submitting a request $p = (p_3, p_2, p_1)$. Under the pessimistic assumption, he expects that Pirate 1 will reject p unless $p_1 > 0$. Hence Pirate 3's optimal proposal is (99, 0, 1). Keeping the pessimistic assumption, Pirate 4's optimal proposal is (99, 0, 1, 0) since to obtain Pirate 2's approval he must offer him at least one coin. Repeating the argument we end up with the subgame perfect-equilibrium payoffs (96, 0, 1, 0, 1, 0, 1, 0, 1, 0) which gives the lion's share to the oldest pirate!

Other subgame perfect equilibria can be found by departing from the pessimistic assumption. For instance, under the optimistic assumption that any pirate approves of an earlier request if he does not expect more by rejecting it, the equilibrium gives <u>all</u> coins to Pirate 10! And there are more equilibria in between these two. (Pirate 10 however, never gets less than 96 coins.)   (See Ruckle [1982].)

Why are Pirates 9, ..., 1 exploited by Pirate 10 at any subgame perfect equilibrium? Because they cannot sustain cooperative promises. For instance, when Pirate 3 proposes (99, 0, 1), Pirate 2 would like to propose to Pirate 1, "Let us reject this proposal; I then will offer you 50 in the next turn." Yet Pirate 1 does not believe the proposal will work in his favor since nothing will prevent Pirate 2 from reneging on his promise <u>and</u> he will have every incentive to do so.

This is the essential feature of subgame perfectness.   It rules out noncredible threats and/or promises that a player would have no interest to carry out afterwards.   In the next chapter we allow for all kinds of threats and promises.   As a consequence, the number of noncooperative equilibria increases dramatically.

4.   EXERCISES

a)   <u>Examples</u>

1)   Consider the following two-person game.

Round 1: Player 1 calls or folds. If he folds, the game is over and the payoffs are (0, 1). If he calls, we go to Round 2. This is a 2 x 2 game played simultaneously (each player being unable to communicate with the other). The payoff matrix is

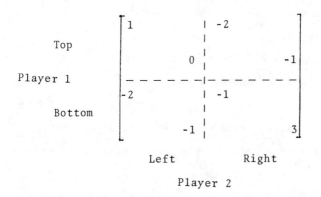

a) Write the normal form of the game and compute the sophisticated equilibrium. Notice that Player 1 wins the competition for the first move of the 2 x 2 game. Interpretation? b) Consider now the new game where in Round 0 Player 2 has the opportunity to fold and guarantee the payoffs (0, 1/2). Write the normal form of the game and compute the sophisticated equilibrium. c) You are Player 1 involved in the game of question b and Player 2 does call. In what sense do you think this is a rational move of your opponent? What do you play?

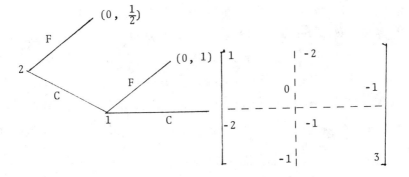

2) First-price auction

Consider the first-price auction game as defined in Example 1, Chapter 3. a) Perform the successive elimination of dominated strategies (two rounds are enough). Notice that the tie-breaking rule makes the computation of X'_i slightly different when $i = 1$ or $i \geq 2$. b) Prove that our game is not dominance solvable, nor is the reduced game inessential. However, in the reduced game, prudent behavior is in general deterministic. Hence, a natural noncooperative outcome of the game results.

3) A game of ascending auction (O'Neill [1985])

A prize of $\$n$, $n \geq 1$, is auctioned off between two players. Each player has a total wealth of $\$r$, $r \geq 1$. The players, starting with Player 1, take turns to raise bids, with a minimal increment $\$1$. Once a player — say i — does not overbid his opponent's last bid, the game is over. Player j gets the prize and <u>both</u> bids are paid.

All variables in this game, n, r as well as all bids, are integer valued. To extract a specific subgame perfect equilibrium we make the following assumption. When the revenue from bidding or not bidding are the same, a player refrains from bidding.

a) Show there is a unique subgame perfect equilibrium under the assumption made above and that Player 1 bids $f(r, n)$, while Player 2 bids nothing (so the game stops after one round). b) Prove that $f(r, n)$ is worth

$$f(r, n) = \text{remainder of the division or } r \text{ by}$$
$$n-1, \text{ if } r/n-1 \text{ is not an integer}$$
$$= n-1, \text{ if } r/n-1 \text{ is an integer}$$

4) In Example 3 (the Steinhaus method to share a cake) the knife is now moved from $x = 1$ to $x = 0$. How is the previous analysis affected?

5) The divide-and-choose method

The cake and player's tastes are the same as in Example 3, but the method is different. Now Player 1 picks a number $x_1 \in [0, 1]$. Next Player 2 chooses either one of $[0, x_1]$ or $[x_1, 1]$ as his share (and Player 1 gets the remaining share). a) State the normal form of the game. b) Prove that each player has a unique prudent strategy and that prudent behavior allows Player 2 to keep all the surplus.

c) Assuming that Player 2, when indifferent about two shares, chooses so as to favor Player 1, prove that the corresponding subgame perfect equilibrium gives all the surplus to Player 1.

d) Compute the optimal utility level of either player acting as a leader. Interpretation?

6) <u>Dividing a shrinking dollar</u> (Rubinstein [1982])

Agents 1, 2, ..., n divide a dollar among themselves according to the following rule:

<u>Step 1</u>: Player 1 proposes a sharing $x^1 = (x_1^1, \ldots, x_n^1)$ where $\sum_{i=1}^{n} x_i^1 = 1$ and $x_i^1 \geq 0$, all $i \in N$. Then agents

2, ..., n each have the option to accept x^1 or reject it. If all agents agree on x^1, it is done. If at least one agent rejects x^2, then we go to Step 2.

<u>Step 2</u>: Player 2 submits a proposal x^2 to the unanimous approval of the other agents. If his request is rejected, we go to Step 3 where agent 3 makes a proposal, and so on. If step n is ever reached, and player n's proposal is rejected, then the whole procedure starts again, with a proposal by Player 1, and so on.

We assume that the initial dollar is depreciated at each period by a discount factor τ, $0 < \tau < 1$. Thus, at period 2, it remains $\delta = 1 - \tau$ dollar to be shared, δ^2 at period 3 and so on. Of course in the event that the division procedure goes on undefinitely, each player ultimately makes a zero profit.

Consider the following simple strategies (much simpler than arbitrary strategies can be in this infinitely long game). Each player, say i, has an acceptance level x_i; he accepts at period t any offer from which he would get at least $\delta^{t-1} \cdot x_i$ and rejects any other. In addition, Player i has a fixed proposal strategy $y^i = (y^i_1, \ldots, y^i_n)$. At time t he proposes the allocation $\delta^{t-1} \cdot y^i$.

Prove that a proper (unique) choice of x^i, y^i, $i = 1, \ldots, n$ makes the corresponding strategy n-tuple a subgame perfect equilibrium of our game. The proof that all subgame perfect equilibria of this game (using even nonsimple strategies) achieve the same sharing of our dollar is more complicated (see Rubinstein [1982]).

b) Abstract games

7) Sophisticated equilibria generalize dominant strategy equilibria

If game G has (at least) one equilibrium in dominating strategies, then G is dominance solvable and its sophisticated equilibrium coincide with its dominant strategy equilibrium. Hint: Use Exercise 7, Chapter 3.

8) Stackelberg equilibrium

Let $G = (X_1, X_2, u_1, u_2)$ be a two-player game where X_i, $i = 1, 2$ are finite and u_i, $i = 1, 2$ are <u>one-to-one</u> on $X_1 \times X_2$. To analyze the behavior of both players when Player 1 is a leader (has the first move) we consider the game

$$L(G, 1) = (X_1, X_2, \tilde{u}_1, \tilde{u}_2) \quad \text{where}$$

$\cdot \xi_2 \in X_2^{X_1}$ is any mapping $x_1 \to \xi_2(x_1)$ from X_1 into X_2.

$\cdot \tilde{u}_i$, $i = 1, 2$, is defined by

$$\tilde{u}_i(x_1, \xi_2) = u_i(x_1, \xi_2(x_1))$$

a) Comment upon this definition. b) Prove that $L(G, 1)$ is dominance solvable and that its sophisticated equilibrium corresponds to the 1-Stackelberg equilibrium of G (Chapter 2, Section 3). c) Compute Player 2's optimal utility level as a leader in $L(G, 1)$. How does this relate to the Wolf-Sheep Lemma in Chapter 2, Section 4?

9) Prove that every finite game in extensive form has at least one subgame perfect equilibrium.

10) <u>Proof of Lemma 1</u> (Gretlein [1983], Rochet [1980]).

Let $G = (X_i, u_i; i = 1, \ldots, n)$ be a fixed finite N-normal form game such that (5) holds. If $B = \underset{i \in N}{X} B_i$ is a rectangular subset of X_N, we denote by G(B) the game $(B_i, u_i; i \in N)$ and by $G(B^t) = (B^t_i, u_i; i \in N)$, the game left from $G(B)$ after t successive eliminations of dominated strategies (Definition 1). For any two rectangular subsets B, C we denote by $C \to B$ the following property.

$$
\text{all } i \in N \left\{
\begin{array}{l}
C_i \subset B_i \\[2ex]
\forall x_i \in B_i \; \exists \; y_i \in C_i : \; \forall x_{-i} \in B_{-i} \; u_i(x_i, x_{-i}) \\[2ex]
= u_i(y_i, x_{-i})
\end{array}
\right.
$$

1) Prove the following implications:

$$
\{C \to B\} \Rightarrow \{C^1 \to B^1\}
$$

2) Deduce that if $C \to B$ holds, we have the following equivalence.

$G(B)$ is dominance solvable $\iff$ $G(C)$ is dominance solvable

If these properties hold, the sophisticated equilibrium payoffs coincide.

We proceed now to prove Lemma 1. We set r to be the cardinality of X_N and assume that Lemma 1 holds for any game whose outcome set has cardinality strictly less than r.

We fix a rectangular subset A of X_N such that (4) holds. If $|A| = r$, there is nothing to prove. Therefore, we assume $|A| \leq r - 1$. We set $B_i = A^1_i \cup X^2_i$ and $X^2 = \underset{i \in N}{X} X^2_i$, and $A^1 = \underset{i \in N}{X} A^1_i$. Observe that

$$
A^1 \subset B \subset A
$$

Thus by the induction assumption we have $G(B)$ is dominance solvable iff $G(A)$ is and their sophisticated equilibrium payoffs coincide.

3) Setting $C = (A^1 \cap X^1) \cup X^2$ prove that $C \to B$.

4) Remark that $X^2 \subset C \subset X^1$ and conclude.

CHAPTER 5.   NASH EQUILIBRIA

1.   THE CONCEPT OF NASH EQUILIBRIUM

Dominant strategy, and prudent and sophisticated behavior can all be sustained independently by the players. Each player on his own, aware only of the normal form of the game, can compute the strategy (or strategies) recommended by either one of these arguments for rationality. Clearly, the timing of various strategic choices plays no role.

On the contrary, Nash equilibria rely on a coordination device (see the agreements and prophecies below) and/or a dynamic scenario (see the Cournot tatonnement in Chapter 6). Some kind of communication among the players is necessary to endow them with mutually consistent beliefs, and/or allow mutual observation of past outcomes. Technically, the argument backing this concept is a stability property (fixed-point argument) instead of a behavioral scenario by which an isolated player decides how to play in view of the mere rule of the game.

Notation

Given an n-player, normal-form game $G = (X_i, u_i, i = 1, ..., n)$, we say that strategy $x_i \in X_i$ is Player i's best reply to strategies $x_{-i} \in X_{-i}$ of the other players if

$$u_i(x_i, x_{-i}) = \max_{y_i \in X_i} u_i(y_i, x_{-i})$$

We denote this property as

$$x_i \in BR_i(x_{-i})$$

where BR_i is a correspondence from X_{-i} into X_i.

Definition 1.

An outcome $x^* \in X_N$ is a Nash equilibrium of G if x^*_i is a best reply to x^*_{-i} for all $i = 1, ..., n$.

$$x^*_i \in BR_i(x^*_{-i}) \qquad all\ i = 1, ..., n$$

We denote by NE(G) the set of Nash equilibria of G.

Our players behave as if they were not aware of their strategic interdependency. When Player i considers a switch from strategy x_i to strategy y_i he does not anticipate a reaction to his move by other players. That is, he does not expect them to change their strategies in response to his own change. This assumption is plausible if the players are many, so that the externality caused by a single deviation to the overall outcome is negligible; alternatively, in the complete ignorance framework, where Player i ignores the utility functions u_j for j in N\{i}, he can acquire information on the u_j by observing the reactions by N\{i} to a switch that would anyway be a profitable one if no reaction occurs (this line of argument is developed in Chapter 6).

Nash equilibria are self-fulfilling prophecies. If every player guesses what strategies are chosen by the others, this guess is consistent with selfish maximization of utilities if and only if all players bet on the same Nash equilibrium. Here, we need an invisible mediator or some theory that pronounces x^*_i the rational strategic choice for Player i. Since every other player can reconstruct the theoretical argument, the Nash equilibrium property is necessary to ensure that our players who are maximizing their utility abide by it.

Nash equilibria also can be viewed as self-enforcing agreements. Think of the players openly discussing the game until they reach a nonbinding agreement to play a certain outcome. The next moment they separate, and all communication between them becomes impossible. Then each player chooses, secretly and ignoring the other actual strategic choices, his actual strategy. He can be faithful to the previous agreement or betray it at no cost. If the agreed outcome is a Nash equilibrium (and only in that case), the agreement is self-enforcing. Assuming that everybody else is loyal, I better be loyal myself (and the more conspicuous my loyalty, the more incentive I give to the others to be loyal).

2. NASH AND SOPHISTICATED EQUILIBRIA

The Nash equilibrium concept generalizes sophisticated equilibrium.

Theorem 1.

Let $G = \{X_i, u_i, i = 1, \ldots, n\}$ be a n-player, normal-form game. Suppose X_i is a compact space and u_i a continuous function on X_N, for all $i = 1, \ldots, n$. Suppose next that in the successive elimination of dominated strategies (Definition 1, Chapter 4), each set X_i^t is compact, $i = 1, \ldots, n$, $t = 1, 2, \ldots$. Then a sophisticated equilibrium of G is a Nash equilibrium as well.

Notice that the topological assumptions in Theorem 1 hold trivially if all X_i are finite, $i = 1, \ldots, n$.

Proof.

Check first the following claim. For all i, t and $x_i \in X_i^t \setminus X_i^{t+1}$ there exists $y_i \in X_i^{t+1}$

such that

$$u_i(x_i, x_{-i}) \leq u_i(y_i, x_{-i}) \qquad \text{all } x_{-i} \; \varepsilon \; X^t_{-i} \qquad (1)$$

Use the argument in Step 1 of the Proof of Lemma 1, Chapter 3. Since X^t_i is assumed to be compact, relation R defined on X^t_i by

$$x_i R y_i \; <=> \; u_i(x_i, x_{-i}) \geq u_i(y_i, x_{-i}) \qquad \text{all } x_{-i} \; \varepsilon \; X^t_{-i}$$

satisfies the assumption of Zorn's Lemma. Therefore, above any $x_i \; \varepsilon \; X^t_i$, there is a maximal element $y_i \; \varepsilon \; X^t_i$ such that $y_i R x_i$. A maximal element of R is but a strategy in X^{t+1}_i.

We prove now Lemma 1 ad absurdum. Pick a sophisticated equilibrium $x^* \; \varepsilon \; X^\infty_N$ and suppose $x^* \notin NE(G)$. Then, for some i and some $x^o_i \; \varepsilon \; X_i$, we have

$$u_i(x^*) < u_i(x^o_i, x^*_{-i}) \qquad (2)$$

Clearly $x^o_i \; \varepsilon \; X^\infty_i$ is impossible since, on X^∞_N, all strategies of Player i are equivalent (to him). Thus there is an integer t such that $x^o_i \; \varepsilon \; X^t_i \backslash X^{t+1}_i$. By (1) we can find $x^1_i \; \varepsilon \; X^{t+1}_i$ such that

$$u_i(x_i, x^*_{-i}) \leq u_i(x^1_i, x^*_{-i})$$

Remember $x^*_{-i} \; \varepsilon \; X^\infty_{-i} \Rightarrow x^*_{-i} \; \varepsilon \; X^t_{-i}$. If we have $x^1_i \; \varepsilon \; X^{t+1}_i \backslash X^{t+2}_i$, repeat the argument to find $x^2_i \; \varepsilon \; X^{t+2}_i$ such that

$$u_i(x^1_i, x^*_{-i}) \leq u_i(x^2_i, x^*_{-i})$$

If $x^1_i \; \varepsilon \; X^{t+2}_i$, set $x^2_i = x^1_i$. By repeating inductively this argument, we construct a sequence $x^p_i \; \varepsilon \; X^{t+p}_i$, $p = 0, 1, \ldots$ such that

$$u_i(x^p_i, x^*_{-i}) \leq u_i(x^{p+1}_i, x^*_{-i}) \text{ all } p = 0, 1, 2, \ldots (3)$$

Take a converging subsequence of x^p_i and call x_i its limit. Since X^q_i are all compact and decreasing w. r. t. q, the strategy x_i belongs to X^q_i for all q. Therefore it belongs to X^∞_i . By the continuity of u_i and inequalities (2) (3) we get $u_i(x^*) < u_i(x_i, x^*_{-i})$ contradicting the dominance solvability of G. QED.

The converse of Theorem 1 is not true. A nondominance solvable game may have (several) Nash equilibria — see, e.g., Example 1 below. A dominance solvable game may have many more Nash equilibria than its sophisticated equilibrium. Our first two examples illustrate these two facts.

Example 1. Where the only Nash equilibrium strategies are dominated.

	Left		Middle		Right	
Top	1		2		3	
		3		0		1
Middle	0		2		0	
		2		2		2
Bottom	3		2		1	
		1		0		3

Left　　　Middle　　　Right

The best reply correspondences are for the row player:
L → B; M → T, M or B; R → T; for the column player: T → L;
M → L, M, or R; B → R. Hence (M, M) is the unique Nash
equilibrium. However, M is dominated by top and bottom for
the row player, and M is dominated by left and right for the
column player. After deleting these dominated strategies,
our players face a two-person, zero-sum game without a value.

Thus, the scenario resulting from the Nash equilibrium
is not always consistent with the, admittedly mild, noncooperative postulate that a player should use an undominated
strategy.

Example 2. <u>Where a Nash equilibrium relies on a noncredible</u>
<u>threat.</u>

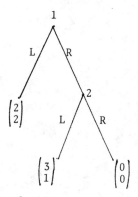

The normal form of this game is

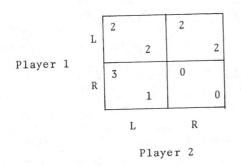

Player 1

	L	R
L	2 2	2 2
R	3 1	0 0

Player 2

Outcome (R, L) is the sophisticated (subgame-perfect) equilibrium. Player 1 wins because he knows that Player 2 will work for him when he has the move. Another Nash equilibrium is (L, R) where Player 2 wins by threatening to play right thus inducing Player 1 to use his prudent strategy to go left. Notice that this is a risky strategy for Player 2 (his only prudent strategy is left).

The situation of Example 2 is quite typical of games in extensive form. While the sophisticated (subgame perfect) equilibrium is generally unique (at least its utility vector is unique: see Theorem 1, Chapter 4) there are many other Nash equilibria with a broad spectrum of associated utility vectors. More examples are given in Exercises 1, 2, 3, 6, and 7.

3. NASH EQUILIBRIA AND TACTICAL COMMUNICATION

Lemma 1.

At any Nash equilibrium of the game $G = (X_i, u_i, i = 1, \ldots, n)$ all players enjoy at least their secure utility *when playing last:*

$$x^* \in NE(G) \Rightarrow u_i(x^*) \geq \inf_{x_{-i} \in X_{-i}} \sup_{x_i \in X_i} u_i(x) \quad all \ i = 1, \ldots, n$$

The obvious proof is omitted.

A consequence is that if a <u>two</u>-person game has a Nash equilibrium, the competition for the second move cannot arise (see Definition 3, Chapter 2). However, the competition for the first move usually arises when several Nash equilibria coexist.

Consider, for instance, the crossing game (Example 1, Chapter 2). There we have two Nash equilibria, namely, (stop, go) and (go, stop). My best reply is to stop if you go, but to go if you stop. The conflictual choice of one of the two Nash equilibria is tantamount to the competition for the first move. A similar configuration arises in the War of Attrition (Example 2, Chapter 2) where the outcomes of the Nash equilibria take the form $(x_1, 0)$, $x_1 \geq 1$ (Player 1 wins) and $(0, x_2)$, $x_2 \geq 1$ (Player 2 wins). Whenever a game has two Nash equilibria which are not Pareto comparable (no single utility-vector Pareto dominates the other, nor are they equal), a conflict akin to the competition for the first move arises. For two-person games a precise formulation of this fact is the subject of Exercise 5 below.

If a game involves several noninterchangeable Nash equilibria (say that (x_1, x_2) and (y_1, y_2) are noninterchangeable

Nash equilibria if (x_1, y_2) and/or (y_1, x_2) is not a Nash equilibrium), players cannot select Nash-equilibrium strategies without some kind of coordination. This should be clear when these Nash equilibria do not yield Pareto-comparable utility vectors. Even when they do, the need for coordination generally persists, as the example shows.

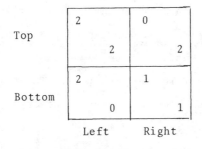

The Nash equilibrium (T, L) is Pareto superior to the Nash equilibrium (B, R). Yet for Player 1, B dominates T and for Player 2, R dominates L. Thus outcome (T, L) is more risky. Each player uses a nonprudent strategy in the belief that the other does the same; but if I suspect that my partner could defect from this cooperative behavior, I better use my own dominating strategy.

The Nash-equilibrium concept does not lead to a rationale of behavior as sophisticated or prudent behavior do. An exception is the class of two-person, zero-sum games, where a Nash equilibrium is just a saddle point, so that Nash

equilibrium strategies coincide with prudent-optimal strategies
(and saddle points are exchangeable; see Theorem 1, Chapter 1).

4. NASH'S THEOREM

One theoretically appealing feature of the Nash-equilibrium
concept is its nice mathematical tractability. Nash's theorem
gives sufficient conditions to guarantee the existence of at
least one Nash equilibrium. We know already of two such
conditions: G is inessential. Then any n-tuple of prudent
strategies is a Nash equilibrium (Theorem 1, Chapter 2). G is
dominance solvable (see Theorem 1 above). However, no tractable
condition is known to imply either inessentiality or dominance
solvability (with the exception of Theorem 3, Chapter 6).

Theorem 2 Nash [1951].

*Suppose that for all $i = 1, \ldots, n$ the strategy set
X_i is convex and compact, and the utility function u_i
is continuous over X and quasiconcave w. r. t. its
variable x_i. Then the game $(X_i, u_i, i = 1, \ldots, n)$ has
at least one Nash equilibrium.*

In this statement we mean that X_i, i = 1, ..., n are convex
and compact subsets of some topological vector space that can
vary with i. By the quasiconcavity of u_i w. r. t. x_i we mean
for all x_i, $y_i \in X_i$, all $x_{-i} \in X_{-i}$, all λ, $0 \le \lambda \le 1$

$$u_i(\lambda x_i + (1 - \lambda)y_i, x_{-i}) \ge \inf\{u_i(x_i, x_{-i}), u_i(y_i, x_{-i})\}$$

A corollary of Nash's theorem is von Neumann's theorem (Theorem 2, Chapter 1). Indeed, a Nash equilibrium in a two-person, zero-sum game is a saddle point. Let us state next a useful variant of Nash's result.

Nash's thoerem for symmetrical games.

Suppose $G = (X_i, u_i, i = 1, ..., n)$ is a symmetrical game $(X_1 = ... = X_n$ and $u_i(x) = u_j(x')$ if x' is deduced from x by exchanging x_i and $x_j)$ with convex compact strategy sets and a utility function u_i which is continuous and quasiconcave in its variable x_i. Then G has (at least) one symmetrical Nash equilibrium x^*:

$$x^*_1 = x^*_2 = ... = x^*_n.$$

Proof.
Both theorems rely on a fixed-point argument.

Kakutani's theorem (Kakutani [1941]).

Let Y be compact and convex (in some topological space) and ϕ be a correspondence associating to each $y \in Y$ a nonempty closed convex subset $\phi(y) \subseteq Y$. Suppose next that ϕ is upper hemicontinuous. Then ϕ has at least one fixed point. There exists $y^* \in Y$, such that $y^* \in \phi(y^*)$.

Consider now a game $G = (X_i, u_i, i = 1, ..., n)$ satisfying the premises of Theorem 2. Set $Y = X_N$ and define ϕ as

$$\phi(x) = BR_1(x_{-1}) \ x \ \ldots \ x \ BR_n(x_{-n}) \qquad \text{for all } x =$$
$$(x_1, \ \ldots, \ x_n) \ \varepsilon \ Y$$

Since X_i are compact and u_i is continuous, $\phi(x)$ is nonempty and closed. Since u_i is quasiconcave in x_i, $BR_i(x_{-i})$ is convex, and so is $\phi(x)$, for all $x \ \varepsilon \ Y$. Finally upper hemicontinuity of ϕ follows from the global continuity of each u_i on Y.

Invoking Kakutani's theorem, there exists an outcome x^* such that $x^* \ \varepsilon \ \phi(x^*)$. This says that x^* is a Nash equilibrium.

To prove the variant of Nash's theorem for symmetrical games, use the following correspondence from X_1 into itself.

$$\psi(x_1) = BR(\tilde{x}_{-1})$$

where $\tilde{x}_j = x_1$, all $j = 2, \ \ldots, \ n$. Once we have a fixed point x_1^* of ψ, the outcome $(x_1^*, \ x_1^*, \ \ldots, \ x_1^*)$ is a symmetrical Nash equilibrium. <u>QED</u>.

Most variants of Nash's existence theorem rely on the convexity of the best reply sets (or some topologically equivalent property, such as acyclicity). An exception is the assumption similar to monotonicity made by Nishimura and Friedman [1981].

To compute all Nash equilibria of a given game G, we need to solve the system

$$u_i(x^*) = \max_{x_i \varepsilon X_i} u_i(x_i, \ x_{-i}^*) \qquad \text{all } i = 1, \ \ldots, \ n$$

Thus, if the u_i are concave in x_i (not just quasiconcave) each global maximization problem is equivalent to a local problem (as we know from convex programming). For instance, if x_i is interior to X_i and u_i is differentiable in x_i, the above system is equivalent to the first-order conditions

$$\frac{\partial u_i}{\partial x_i} (x^*) = 0 \qquad \text{all } i = 1, \ldots, n \quad (4)$$

System (4) is expected to have isolated solutions, as its number of independent equations equals the dimension of X_N. This implies that <u>Nash equilibria outcomes are not, in general, Pareto optimal</u> (for a precise genericity result, see Grote [1974]).

Guaranteeing <u>uniqueness</u> of the Nash equilibrium is difficult. The only available results derive uniqueness of the Nash equilibria from the <u>stability</u> of some adjustment process, such as Cournot's tatonnement, to which Chapter 6 is devoted. The most-useful uniqueness result actually relies on a continuous process (differential equation). We state it, without proof, in the simple framework of games over the unit cube.

Theorem 3 Rosen [1965].

Suppose $X_i = [a_i, b_i]$ is a compact real interval, for all i, and u_i is a C^2 smooth function on X_N satisfying

$$\frac{\partial^2 u_i}{\partial x_i^2} (x) < 0 \qquad \text{all } x \in X_N$$

Then, denote by Q the $n \times n$ matrix with (i, j) entry $(\partial^2 u_i)/(\partial x_i \partial x_j)$. If $Q + {}^t Q$ is negative definite for all $x \in X_N$, then the game $(X_1, \ldots, X_n; u_1, \ldots, u_n)$ has a unique Nash equilibrium.

For instance, in a two-person game, the negative definiteness assumption amounts to

$$\frac{\partial^2 u_i}{\partial x_i^2} (x) < 0 \qquad \text{all } i = 1, 2$$

all $x \in X_{12}$

$$\left| \frac{\partial^2 u_1}{\partial x_1 \partial x_2} + \frac{\partial^2 u_2}{\partial x_1 \partial x_2} \right| \leq 2 \left| \frac{\partial^2 u_1}{\partial x_1^2} \cdot \frac{\partial^2 u_2}{\partial x_2^2} \right|^{1/2}$$

Example 3. <u>Cournot's quantity-setting oligopoly.</u>

The n players are n firms controlling respectively the supply $x_1, \ldots, x_n$ of some satiable good. Given the total supply $\bar{x} = x_1 + \ldots + x_n$, the price is given by the inverse demand function $p(\bar{x})$. We assume that p is continuous over $[0, +\infty[$, and there is a satiation level $S > 0$ such that for $0 \leq y \leq S$, $p(y)$ is strictly decreasing, concave and twice differentiable (hence $p'(y) \leq 0$ and $p''(y) \leq 0$). For $S \leq y$, $p(y) = 0$. Firm i has a maximal capacity γ_i and its cost function c_i is continuous, strictly increasing and convex over $[0, \gamma_i]$ (hence $c'_i \geq 0$, $c''_i \geq 0$). Hence, the normal form game

$$X_i = [0, \gamma_i] \quad u_i(x) = x_i p(\bar{x}) - c_i(x_i) \quad \text{all } i = 1, \ldots, n$$

where u_i is continuous and quasiconcave w. r. t. x_i. Thus a Nash equilibrium x^* exists. Clearly $\bar{x}^* < S$. Thus, x^* is determined by the first-order conditions:

$$x_i^* \, p'(\bar{x}^*) + p(\bar{x}^*) = c_i'(x_i^*) \qquad i = 1, \ldots, n$$

At any time during the solution of this system, the overall supply $\bar{x}^*$ exceeds the supply that would maximize joint profit (when properly allocated among firms). To see this, set $\pi(x) = \bar{x}p(\bar{x}) - \sum_{i=1}^{n} c_i(x_i)$ to be the joint profit at outcome x, and compute

$$\frac{\partial \pi}{\partial x_i}(x^*) = (\bar{x} - x_i) \cdot p'(\bar{x}^*) < 0$$

So reducing any one of the individual supplies increases joint profit.

Remark

In the model described above existence of a Nash equilibrium is guaranteed with much weaker assumptions on the cost functions c_i. It is enough that c_i is strictly increasing over $[0, \gamma_i]$ and s. t.

$$\lim_{x_i \to 0} \frac{c_i(x_i) - c_i(0)}{x_i} > 0$$

(See Novshek, [1983] and Exercise 10 for a particular case of this result). More games modelling oligopolistic competition are proposed in the Exercises 8, 9, and 10; see also Exercises 2 and 3 of Chapter 6.

5. EXERCISES

a) Games in normal form

1) Auction dollar game

Consider the two-person game of Example 2, Chapter 2. Compute the best reply correspondences and the Nash equilibria. Are they Pareto-optimal outcomes?

2) First-price auction

Consider the first-price auction game (defined in Example 1, Chapter 3; see also Exercise 2, Chapter 4) and suppose the values of the object to the players are such that

$$a_1 > a_2 \geq a_j \qquad\qquad \text{all } j \geq 2$$

Prove that the Nash equilibria are such that Player 1 gets the object at some price p, $a_2 \leq p \leq a_1$. Prove that the range of p is the whole interval $[a_2, a_1]$.

3) Second-price auction

In the second-price auction game (Example 1, Chapter 3) show that Nash equilibria are very numerous. More precisely, for any player i and any price p, such that $0 < p \leq a_i$, prove the existence of at least one Nash equilibrium where Player i gets the object and pays p for it.

4) Give an example of a 2 x 2 game (two players, two strategies each) where each player has a unique prudent strategy x_i^* ;

where $x^* = (x_1^*, x_2^*)$ is a Nash equilibrium and a Pareto-optimal outcome; and x^* is the only outcome such that $\alpha_i \leq u_i(x^*)$ i = 1, 2; yet the game is <u>not</u> inessential since $\alpha_i < u_i(x^*)$ for i = 1, 2.

5) Suppose the two-person game $G = (X_1, X_2, u_1, u_2)$ has two Pareto-optimal Nash equilibria x, y, with distinct utility vectors $(u_1(x), u_2(x)) \neq (u_1(y), u_2(y))$.

Prove that the competition for the first move arises in G.

b) <u>Games in extensive form</u>.

6) Consider the following game

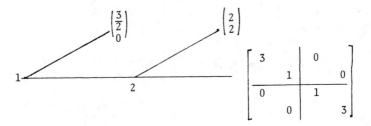

Write the normal form of the game. Compute its sophisticated equilibrium x^*. Prove that our game has one more Nash equilibrium y^* and that y^* Pareto dominates x^*. Explain how the players can enforce y^*.

7) <u>Voting by veto</u>

Consider the game of voting by veto described in Example 4, Chapter 4. Which candidates can be elected at some Nash equilibrium? Answer the same question in the new game where

Player 1's preferences are now

$$u'_1 (c) < u'_1 (b) < u'_1 (d) < u'_1 (a)$$

c) <u>Oligopoly models</u>

8) <u>Cournot duopoly with fixed costs</u>

This is a quantity-setting oligopoly where both firms
supply respectively the quantities x_1, x_2 and the resulting
price (at which all supply is sold) is $p = 1 - x_1 - x_2$.
Production involves consideration of a fixed cost a > 0 to
produce any positive amount of the good, and no variable cost.
Hence the game

$$X_1 = X_2 = [0, 1]$$

$$u_i(x_1, x_2) = x_i(1 - x_1 - x_2) - a \qquad \text{if } x_i > 0$$

$$= 0 \qquad \text{if } x_i = 0$$

We discuss this game with respect to the parameter a.
a) Show that for a small (0 < a < α for some α to be computed)
the game is strategically similar to the costless game (a = 0).
There is a unique Nash equilibrium; it is also the sophisticated
equilibrium of the game. b) For a not-too-small and not-too-big
(α ≤ a ≤ β for some β to be computed) there are three equilibria
(in pure strategies). One of them is the sophisticated
equilibrium. c) For a large (β < a ≤ 1/4) there are two
asymmetrical equilibria in pure strategies, where only one

firm is active. In this case there is also a symmetrical
equilibrium in mixed strategies (see Chapter 7 below).

9) Duopoly à la Bertrand

Two firms sell the same good. Their strategy is to establish
a price $x_i \geq 0$. If $x_i < x_j$, firm i must satisfy all the demand
$D(x_i) = 300 - 5x_i$ and firm j sells nothing. If $x_1 = x_2$, the
firms share equally the demand $D(x_1)$. a) Assume first that
the cost function is $C(q) = 10q$ for both players. Write the
normal form of the game, with strategy sets

$$X_1 = X_2 = [0, 60]$$

Show that there is a unique Nash equilibrium: compare the
corresponding profits to the minimal guaranteed profits.
b) Assume next that the cost functions differ

$$C_1(q_1) = 10q_1 \quad C_2(q_2) = 20q_2$$

Prove that, strictly speaking, there is no Nash equilibrium
in this game. However, define the ε Nash equilibrium as an
outcome where no player, by a unilateral deviation, can
improve his utility by more than ε. Then prove that there
are infinitely many ε Nash equilibria in this game and describe
them. c) In the game in b) perform two rounds of elimination
of dominated strategies; show that the reduced game is
inessential and its equilibrium corresponds to the ε Nash
equilibrium most favorable to Player 1.

10) <u>Existence of a Nash equilibrium in a quantity-setting duopoly</u> (Fraysse)

The inverse demand function is continuous over $[0, +\infty[$, strictly decreasing and concave on $[0, S]$, and zero after S. The two cost functions c_i, $i = 1, 2$ are continuous on $[0, +\infty]$ and such that:

$$c_i(0) = 0$$

$$c_i(x_i) \geq 1 \qquad\qquad \text{for } x_i > 0$$

a) Compute the guaranteed utility level to each player.

b) Show that the best-reply correspondences BR_i are nonincreasing in the following sense:

$$\{x_j < y_j, \; x_i \in BR_i(x_j), \; y_i \in BR_i(y_j)\} \implies \{y_i \leq x_i\}$$

c) Deduce the existence of a Nash equilibrium by picking two single-valued selections b_1, b_2 of BR_1, BR_2 respectively and showing that $b_1 \circ b_2$ has a fixed point on $[0, S]$.

CHAPTER 6.   STABILITY OF NASH EQUILIBRIA

1.   GLOBAL STABILITY

The rationality argument underlying the Nash-equilibrium
concept is akin to the price-taking assumption of perfect
competition.   I do not anticipate any reaction from other
players to a switch of my strategic choice.   This is plausible
if there are many players (so that the influence of each
player is negligibly small), or if the myopic players
completely ignore each other's preferences, so that assuming
no reaction is a coarse but simple starting assumption.

The Cournot tatonnement is a dynamic process where each
player adjusts his strategy optimally under the myopic
assumption that others will not move anymore.   (This assumption

is actually contradicted after each observation!) If this process converges, we have reached a Nash equilibrium. This dynamical system plays the role of a coordination device among players who do not need to communicate. As it could start from any outcome, it is a plausible scenario to reach a Nash equilibrium only if it is a stable system. From any initial position it eventually converges to some Nash equilibrium.

Example 1. Binary choices with externalities.

Here we assume a large number of identical players each taking a $\{0, 1\}$-binary decision. The general normal form is thus described by two continuous functions a, b both defined on $[0, 1]$:

$a(t)$ is the utility of a player using $x_i = 1$ given that a fraction t of players use strategy 1

$b(t)$ is the utility of a player using $x_i = 0$ given that a fraction t of players use strategy 1

Hence the normal form game

$$X_1 = \ldots = X_n = \{0, 1\}, \text{ n large}$$

$$u_i(x_i, x_{-i}) = a(t) \text{ if } x_i = 1$$
$$= b(t) \text{ if } x_i = 0$$

where $t = \frac{1}{n} \sum_{j=1}^{n} x_j$

Schelling [1971] gives many striking examples of such games, such as requiring players to decide to use a private automobile ($x_i = 0$) or to take the bus ($x_i = 1$), and $a(.)b(.)$ take into account traffic's congestion; or to decide whether to take a

risky vaccination ($x_i = 1$) or not ($x_i = 0$) against a specific
disease. Both examples correspond to functions a(.)b(.)
shaped as in Figure 1.

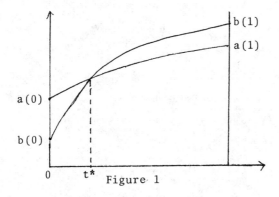

Figure 1

Using strategy 1 is beneficial to both types of players
at any configuration (a, b are both increasing in t). When
everyone is selfish (strategy 0), it is better to be cooperative. Thus, a(0) > b(0). (For example, the traffic is too
heavy and buses have special lanes; or the risk of getting
the disease is so high that you prefer to take the chance of
the inoculation.) When everyone is cooperative (strategy 1)
you are better off if you decide to be selfish. Thus, a(1) <
b(1). (In light traffic, you prefer to use your own car; if
virtually everyone is vaccinated, your exposure to the disease
is nearly impossible, so you more or less safely can avoid
the inoculation.)

On Figure 1, the level t* (a(t*) = b(t*)) corresponds
to outcomes of Nash equilibria. If a single player switches

from $x_i = 0$ to $x_i = 1$, or the other way around, this does not significantly affect the fraction t^*. Thus, $a(t^*) = b(t^*)$ means that no individual has an incentive to switch. Suppose next that a significant fraction of players switch from $x_i = 0$ to $x_i = 1$ so as to change the fraction t to $t^* + \varepsilon$, ε being small. This creates an incentive to switch back as $a(t^* + \varepsilon) < b(t^* + \varepsilon)$. Similarly a switch from $x_i = 1$ to $x_i = 0$, moving t to $t^* - \varepsilon$, yields $b(t^* - \varepsilon) < a(t^* - \varepsilon)$. As a result, some players will switch from $x_i = 0$ to $x_i = 1$. In terms of Definition 1 below, any outcome x^* with $t^* = 1/n \sum_{i=1}^{n} x_i^*$ is a stable Nash equilibrium.

But consider a pair of curves such as those in Figure 2.

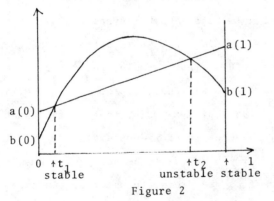

Figure 2

Now we have three values t_1, t_2, 1 at which the curves intersect. At each such value we have a Nash equilibrium. Yet only two of them, namely t_1 and 1, are stable. Therefore, from t_2, any switch to $t_2 + \varepsilon$ (respectively to $t - \varepsilon$) creates an incentive to make more switches in the same direction as $b(t_2 + \varepsilon) < a(t_2 + \varepsilon)$ (respectively $a(t_2 - \varepsilon) < b(t_2 - \varepsilon)$). We say that an outcome x* with $t_2 = 1/n \sum_{i=1}^{n} x^*_i$ is a (locally) unstable Nash equilibrium.

Remark 1.

In Example 1, the assumption that n is large was used in a loose way to derive statements such as t is not affected by an individual switch. The same qualitative results, however, can be obtained by a rigorous and cumbersome argument.

In the remainder of Chapter 6, we return to games with a fixed and finite number of players.

Definition 1.

Let X_i be endowed with some topology for all $i \in N$.
Let $G = \{X_i, u_i; i = 1, \ldots, n\}$ be an N normal form
game. We assume that every player has a unique
best-reply strategy to any fixed strategies by the
other agents:

$$for\ all\ i \in N,\ all\ x_{-i} \in X_{-i}:$$

$$BR_i(x_{-i})\ is\ a\ singleton\ denoted\ r_i(x_{-i}) \in X_i$$

(1)

To any outcome $x^0 \in X_N$ we associate the (simultaneous)
Cournot tatonnement starting at x^0, namely, the following
sequence $x^0, x^1, \ldots, x^t, \ldots$ of X_N:

$$x^t_i = r_i(x^{t-1}_{-i}) \qquad \textit{all } i \in N \textit{ and}$$
$$\textit{all } t = 1, 2, \ldots \qquad (2)$$

We say that a Nash equilibrium outcome x* is stable in G if for any initial position $x^0 \in X_N$ the Cournot tatonnement starting at x^0 converges to x*.

Notice that a stable NE outcome of G necessarily is the unique NE of G, for if the initial position is a NE, the Cournot tatonnement is a constant sequence.

Example 2. A Cournot's quantity setting duopoly.

Consider a quantity setting duopoly a la Cournot (see Example 8, Chapter 5) where

$$p(\bar{x}) = 1 - \bar{x} \qquad\qquad \gamma_i = \frac{1}{2}$$

and the marginal production cost is constant:

$$c_i(x) = \frac{1}{2} x \qquad\qquad i = 1, 2$$

The normal form is

$$X_1 = X_2 = [0, \tfrac{1}{2}], \quad u_i(x_1, x_2) = x_i \cdot (1 - x_1 - x_2) - \frac{1}{2} x_i$$

$$i = 1, 2$$

with best-reply functions

$$r_i(x_j) = \frac{1}{4} - \frac{1}{2} x_j \qquad\qquad 0 \le x_j \le \frac{1}{2}$$

The unique Nash equilibrium (1/6, 1/6) is easily seen to

be stable in Figure 3. Notice that the simultaneous Cournot
tatonnement $(x_1, x_2) \rightarrow (r_1(x_2), r_2(x_1)) \rightarrow (r_1 r_2(x_1), r_2 r_1(x_2))$
is described by two sequences

$$x_1 \rightarrow r_2(x_1) \rightarrow r_1 r_2(x_1) \rightarrow \ldots$$

and
$$(2)$$

$$x_2 \rightarrow r_1(x_2) \rightarrow r_2 r_1(x_2) \rightarrow \ldots$$

allowing an easy graphical representation of stability.

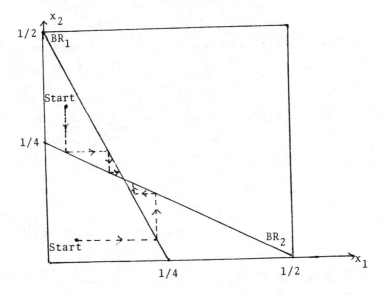

Figure 3

Suppose now that the production cost shows increasing returns to scale as in

$$c_i(x) = \frac{1}{2} x - \frac{3}{4} x^2 \qquad i = 1, 2$$

Best-reply functions are computed:

$$r_i(x_j) = \frac{1}{2} \qquad \text{if } 0 \le x_j \le \frac{1}{4}$$

$$= 1 - 2x_j \qquad \text{if } \frac{1}{4} \le x_j \le \frac{1}{2}$$

We have three Nash equilibria $(1/3, 1/3)$ $(1/2, 0)$ and $(0, 1/2)$. No one of them is stable since the equilibrium is not unique. The Cournot tatonnement does not always converge. For instance, starting from x^o with $0 \le x^o_i \le 1/4$, we have $x^1 = (1/2, 1/2)$, $x^2 = (0, 0)$, $x^3 = x^1$, $x^4 = x^2$, ... However, starting nearby $(0, 1/2)$ (respectively $(1/2, 0)$) we stay nearby, and converge back to $(0, 1/2)$, (respectively $(1/2, 0)$) in finitely many steps. We say that $(0, 1/2)$, $(1/2, 0)$ are two locally stable equilibria, while $(1/3, 1/3)$ is unstable even locally.

The application of the concept of stable Nash equilibria is limited by two technical difficulties. First, our assumption that any best reply is single valued for all players and all outcomes is a strong one. Second, the choice of the simultaneous version of Cournot's tatonnement (all players react today to yesterday's choices) rather than some

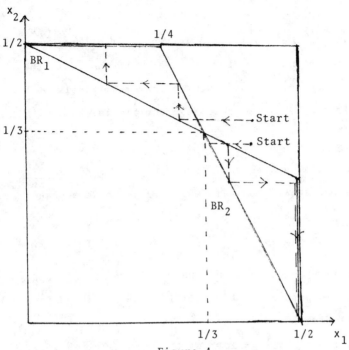

Figure 4

sequential version (only one player reacts at a time) is
arbitrary. In games with at least three players, both
notions of stability do not coincide. (See Exercise 4.)

Last but not least, sufficient conditions for a Nash
equilibrium to be stable are quite restrictive. (See
Theorem 3 below.)

Most of these difficulties (except the discrepancy between
simultaneous and sequential versions of the tatonnement) disappear when we turn to local stability.

134

2. LOCAL STABILITY

We weaken the stability condition by requiring only that
the Cournot tatonnement starting nearby x* should in the
limit approach x*. We are able to characterize almost
completely the locally stable Nash equilibria.

Definition 2 (same notation as in Definition 1).

We say that a Nash equilibrium x is locally stable in G
if there exists for all i = 1, ..., n a neighborhood V_i
of x_i such that assumption (1) holds on V_N and x* is
stable in the restricted game (V_i, u_i, i = 1, ..., n).*

To derive a computational characterization of local
stability, we assume that for all i = 1, ..., n, X_i is a subset
of an euclidian space E_i, and we fix a Nash equilibrium x*
such that x_i^* is an interior point of X_i, all i = 1, ..., n.

We assume, moreover, that the utility functions u_i are
twice continuously differentiable in a neighborhood of x_i^*
and that the second derivative $(\partial^2 u_i)/(\partial x_i^2)$ is a definite
negative operator at x*. Therefore (1) holds in a suitable
neighborhood of x*.

We define a linear operator T from $E_N = \prod_{i \in N} E_i$ into
itself. For all e ε E_N

$$T_i(e) = \sum_{\substack{j=1,...,n \\ j \neq i}} \left(\frac{\partial^2 u_i}{\partial x_i^2}\right)^{-1} \circ \left(\frac{\partial^2 u_i}{\partial x_i \partial x_j}\right) (e_j) \qquad (3)$$

where all the derivatives are taken at x*.

Theorem 1

Suppose that the spectral radius of T is strictly less than one. Then x* is a locally stable Nash equilibrium. Suppose that x* is a locally stable NE outcome. Then the spectral radius of T is less than or equal to one.

Recall that the spectral radius of T is the maximal modulus of its eigenvalues.

Proof.

Since $(\partial u_i)/(\partial x_i)$ is C^1 differentiable and $(\partial^2 u_i)/(\partial x_i^2)$ is nonsingular at x*, the implicit function theorem shows that r_i is, locally at x*, a C^1 differentiable function from x_{-i} into x_i. Henceforth system (2) can be written locally as

$$x^t = f(x^{t-1}) \tag{4}$$

where f is C^1 differentiable from E_N into itself.

By assumption x* is a fixed point of f. Thus, by an elementary result in dynamic systems (see, for example, Ortega Rheinboldt [1970]) we know that x* is a locally stable solution of (4) if the modulus of all eigenvalues of f'(x*) is strictly less than one. Conversely, if at least one eigenvalue of f'(x*) has a modulus strictly above one, then x* is not locally stable. We let the reader check that T = f'(x*). QED.

Corollary.

Suppose n = 2 and X_1, X_2 are one dimensional. Let x* be a NE of $G = (X_1, X_2, u_1, u_2)$ such that x_i^* is an *interior point of X_i, u_i is C^2 differentiable in a neighborhood of x^*, $(\partial^2 u_i)/(\partial x_i^2)$ $(x^*) < 0$. Then we have*

$$\left| \frac{\partial^2 u_1}{\partial x_1 \partial x_2} \cdot \frac{\partial^2 u_2}{\partial x_1 \partial x_2} \right| < \left| \frac{\partial^2 u_1}{\partial x_1^2} \cdot \frac{\partial^2 u_2}{\partial x_2^2} \right| \Rightarrow \quad x^* \text{ is locally stable}$$

$$\left| \frac{\partial^2 u_1}{\partial x_1 \partial x_2} \cdot \frac{\partial^2 u_2}{\partial x_1 \partial x_2} \right| > \left| \frac{\partial^2 u_1}{\partial x_1^2} \cdot \frac{\partial^2 u_2}{\partial x_2^2} \right| \Rightarrow \quad x^* \text{ is not locally stable}$$

(5)

(where all these derivatives are taken at x^*).

Under the assumptions of the corollary, the best-reply sets BR_i are two C^1 curves that intersect at x^*. The inequalities in (5) simply compare the modulus of the slopes

$$s_i = -\frac{\partial^2 u_i}{\partial x_i \partial x_1} \bigg/ \frac{\partial^2 u_i}{\partial x_i \partial x_2}$$

of BR_i, $i = 1, 2$:

$$|s_1| > |s_2| \Rightarrow \quad x^* \text{ is locally stable}$$

$$|s_1| < |s_2| \Rightarrow \quad x^* \text{ is not locally stable}$$

Example 3. <u>A game where all Nash equilibria are locally unstable.</u>

Consider a game on the unit square ($X_1 = X_2 = [0, 1]$) with the following best-reply functions

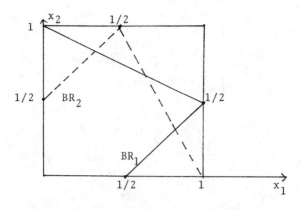

Figure 5

None of the equilibria is stable since the modulus
of the slope to BR_2 always exceeds that of BR_1 when they intersect. At the symmetrical Nash equilibrium (2/3, 2/3), sequences
(2) follow the familiar exploding cobweb pattern. The very
complex structure of the global Cournot tatonnement in such a
game is analyzed by Rand [1978].

Our next example is computational.

Example 4.   Stability in Cournot's quantity-setting oligopoly.

In the general model of Example 3, Chapter 5 we compute
the matrix of the operator T:

$$t_{ij} = - \frac{x_i^* \, p''(\overline{x}^*) + p'(\overline{x}^*)}{x_i^* \, p''(\overline{x}^*) + 2p'(\overline{x}^*) - c_i''(x_i^*)}$$

As t_{ij} is independent of j denote $t_{ij} = -t_i$. Notice that
$0 < t_i < 1$. Next we observe that the determinant of the matrix

$$T - \lambda I = \begin{bmatrix} -\lambda & -t_1 & \cdots & -t_1 \\ -t_2 & -\lambda & \cdots & -t_2 \\ -t_n & \cdots & -t_n & -\lambda \end{bmatrix}$$

is worth

$$(-1)^n \det(T - \lambda I) = \prod_{i=1}^{n} (\lambda - t_i) + \sum_{i=1}^{n} t_i \{ \prod_{j \neq i}^{n} (\lambda - t_j) \}$$

Hence, assuming first that $t_1, \ldots, t_n$ are all distincts, the eigenvalues of T are the solutions of

$$\sum_{i=1}^{n} \frac{t_i}{t_i - \lambda} = 1$$

The equation above has n distinct real roots. They are all in $[-1, +1]$ iff $\sum_{i=1}^{n} \frac{t_i}{t_i + 1} < 1$, which holds true, for instance, if $(t_i)/(t_i + 1) < 1/n$ for all i. That is to say

$$(n - 2)x_i^* |p''(\overline{x}^*)| + (n - 3)|p'(\overline{x}^*)| < c_i'' (x_i^*)$$

(6)

all $i = 1, \ldots, n.$

If some of the t_i coincide, all multiple values of t_i are eigenvalues and the rest are solutions of

$$\sum_{i=1}^{n^*} \frac{m_i t_i}{t_i - \lambda} = 1$$

where $t_1, \ldots, t_{n}^*$ are all distinct and t_i has multiplicity

m_i. Thus the statement that all eigenvalues of T are in
$[-1, +1]$ iff $\sum\limits_{i=1}^{n} (t_i)/(t_i + 1) < 1$ still holds.

3. DOMINANCE SOLVABILITY AND STABILITY

The aim of this section is to exhibit deep mathematical connections between dominance solvability and the stability of the Cournot tatonnement. This attempt might be surprising because the two strategic scenarios justifying these two concepts are so widely different. In one, completely informed players are involved in a one-shot game. In the other myopic players are involved in a dynamic scenario requiring their constant readjustment to a changing environment. Of course, whenever a normal form game has a globally stable Nash equilibrium which is its sophisticated equilibrium as well, the argument for seeing it as the plausible noncooperative outcome of the game is very strong. Therefore Theorem 3 below delineates an admittedly small class of games where noncooperation clearly determines the solution of the game.

Throughout this section, we assume that strategy sets X_i are compact, utility functions are continuous on X_N, and the best replies are single-valued (property (1)). Under the latter assumption the game $G = (X_i, u_i, i = 1, \ldots, n)$ is dominance solvable if and only if the set $X_N^{\infty} = \bigcap\limits_{t=1}^{\infty} X_N^t$ is a singleton x^* (notations as in Definition 1, Chapter 4). In that case x^* is the unique Nash equilibrium of G. Namely, for any $x \in NE(G)$, $x \in X_N^o$ and $x \in X_N^t$ implies $x \in X_N^{t+1}$ (because x_i is the unique best reply to $x_{-i} \in X_{-i}^t$ it cannot be dominated in X_i^t). Hence X_N^∞ contains all Nash equilibria of G. Conversely we know (Theorem 1, Chapter 4) that the sophisticated equilibrium is in NE(G).

Theorem 2.

If G is dominance solvable, its sophisticated equilibrium is also a (globally) stable Nash equilibrium.

Proof.

Denote by R the set of rectangular subsets of X_N. Thus a generic element of R is $R = R_1 \times \ldots \times R_n$ where R_i is a subset of X_i. To simplify notations we denote the set of Player i's undominated strategies in the restriction of our game to R simply by $D_i(R)$.

For any $R \in R$, we prove first the following inclusion.

$$r_i(R_{-i}) \cap R_i \subset D_i(R) \qquad \text{all } i = 1, \ldots, n \quad (7$$

Pick $x_i \in R_i$ such that for some $x_{-i} \in R_{-i}$, $x_i = r_i(x_{-i})$. For any $y_i \in R_i$, uniqueness of the best reply implies

$$u_i(x_i, x_{-i}) \leq u_i(y_i, x_{-i}) \Rightarrow x_i = y_i$$

Hence, x_i cannot be a dominated strategy.

Let (R^t) be the successive elimination of dominated strategies of G. From (7) we get

$$r_i(R^o_{-i}) = r_i(R^o_{-i}) \cap R^o_i \subset R^1_i$$

By induction, suppose $r_i(R^{t-1}_{-i}) \subset R^t_i$. Then $r_i(R^t_{-i}) \subset R^t_i$ (since R^t_j is decreasing with respect to t) so that

$$r_i(R^t_{-i}) = r_i(R^t_{-i}) \cap R^t_i \subset R^{t+1}_i \tag{8}$$

Hence (8) holds for all $t \geq 0$ and all $i \in N$.

Define now the mapping Δ from R into itself:

$$\text{For all } R \in R, \quad \Delta(R) = \prod_{i \in N} r_i(R_{-i})$$

Relations (8) are then rewritten as $\Delta(R^t) \subset R^{t+1}$, all $t \geq 0$. Since Δ is monotonic $(R \subset R' \Rightarrow \Delta(R) \subset \Delta(R'))$ and $R^o = X$, this implies: $\Delta^t(X) \subset R^t$, all $t \geq 0$. $\tag{9}$
Define now the mapping r from X_N into itself.

$$r(x) = (r_1(x_{-1}), \ldots, r_n(x_{-n}))$$

Next denote $B(R) = r(R)$ for all $R \in R$. Then $B(R) \subset \Delta(R)$, but the inverse inclusion is not true in general for $n \geq 3$. (As an exercise, explain why.) From (9) we deduce now $B^t(X) \subset R^t$, all $t \geq 0$. In particular

$$\bigcap_{t \in N} R^t = \{x^*\} \Rightarrow \bigcap_{t \in N} B^t(X) = \{x^*\}$$

Taking any starting point x^o the Cournot tatonnement $x^{t+1} = B(x^t)$ is such that $x^t \in B^t(X)$. Since $B^t(X)$, $t = 1, 2, \ldots$ form a decreasing sequence of compact sets, this implies $\lim_{t \to +\infty} x^t = x^*$ as was to be proved. QED.

The converse of Theorem 1 does not hold in general (see Moulin [1984]). We mention now two computational results about dominance solvability. The first result gives a sufficient condition for the dominance solvability of a game on the unit cube.

Theorem 3 Gabay-Moulin [1980], Moulin [1984].

Suppose $X_i = [0, 1]$ all $i = 1, \ldots, n$ and u_i is twice differentiable on X_N with

$$\frac{\partial^2 u_i}{\partial x_i^2} (x) < 0 \qquad\qquad \text{all } x \in X \text{ and}$$
$$\text{all } i = 1, \ldots, n$$

In particular, u_i is strictly concave in x_i and its best reply is single valued. Suppose moreover that the following inequalities hold true.

$$\sum_{\substack{j=1, \ldots, n \\ j \neq i}} \left| \frac{\partial^2 u_i}{\partial x_i \partial x_j} (x) \right| < \left| \frac{\partial^2 u_i}{\partial x_i^2} (x) \right| \qquad \begin{array}{l} \text{all } i = 1, \ldots, n \\ \text{all } x \in X. \end{array} \qquad (10)$$

Then G is dominance solvable. Hence, it is globally stable with a unique Nash equilibrium.

It is worth mentioning that under the premises of Theorem 3 any sequential Cournot tatonnement (where players take turns to adjust their strategies) is globally stable in G as well.

For instance, inequalities (10) applied to Cournot's oligopoly model (Example 4 above) yield system (6) at any outcome x! Our last result characterizes a local version of dominance solvability and parallels the characterization of local stability (Theorem 1).

Definition 3 same notations as in Definition 2.

We say that a Nash equilibrium x* is locally dominance solvable, if there exists for all i = 1, ..., n a compact neighborhood V_i of x_i such that the restricted game $(V_i, u_i, i = 1, ..., n)$ is dominance solvable with unique sophisticated equilibrium x*.

Our computational result assumes that X_i is a onedimensional interval, with x_i^* as an interior point. We also assume (as in Theorem 1) that $(\partial^2 u_i)/(\partial x_i^2)$ is defined around x* and strictly negative. The matrix T with off-diagonal elements

$$t_{ij} = (\frac{\partial^2 u_i}{\partial x_i \partial x_j})/\frac{\partial^2 u_i}{\partial x_i^2})(x^*)$$

is now replaced by the matrix $|T|$ with off-diagonal elements $|t_{ij}|$ (and zero on the diagonal).

Theorem 4 Moulin [1984].

Suppose that the spectral radius of $|T|$ is strictly less than one. Then x* is dominance solvable locally. Conversely, suppose that x* is a locally dominance solvable Nash equilibrium. Then the spectral radius of $|T|$ is less than or equal to one.

Since the spectral radius of $|T|$ is greater than or equal to that of T, we deduce that local dominance solvability is, in general, a stronger statement than local stability. However, if the sign of the entries t_{ij} of T does not vary (as in Example 4 above) both statements are equivalent (except perhaps when the spectral radius of T is 1).

4. EXERCISES

1) Binary choices with externalities

Analyze the game of Example 1, in the following cases:
a) $a(t) = t$; $b(t) = t + 1/2$; $0 \le t \le 1$. b) $a(t) = (t - 1/3)^2$ $b(t) = (t - 2/3)^2$; $0 \le t \le 1$. In both cases, determine the noncooperative equilibria of the game (dominant strategies, sophisticated, Nash stable, and Nash unstable) and find if they are Pareto optimal. Comment about what noncooperative behavior seems reasonable.

2) Quantity-setting oligopoly with limited entry

Ten identical firms produce the same good with the following cost function.

$$c(x_i) = 9x_i + x_i(x_i - 100)^2$$

The inverse demand dunction is

$$p = 50 - \frac{1}{10} x \qquad \text{if } x \le 500$$

$$= 0 \qquad \text{for } x \ge 500$$

a) Write the normal form of the game (with unlimited production capacity). Check that the Nash theorem does <u>not</u> apply. b) Prove that in any Nash equilibrium at most four firms can be active (i.e., have $x_i > 0$), then compute all the Nash equilibria of the game. <u>Hint</u>: at any Nash equilibrium the price cannot be smaller than the average cost. c) Are the Nash equilibria of the game locally stable?

3) <u>The auto dealer game</u> (Case [1979])

The n players are n auto dealers facing a constant overall fixed demand D. Let the strategy x_i be the number of cars that Dealer i keeps on hand. Assuming that each dealer has the same number of visitors per unit of time, Dealer i faces the demand flow

$$D \cdot \frac{x_i}{\bar{x}} \qquad \text{where } \bar{x} = x_1 + \ldots + x_n$$

Let P_i be his unit profit and C_i his unit cost of storing (per unit of time). Then the following normal-form game emerges.

$$X_i = [0, +\infty[, \quad u_i(x) = D \cdot P_i \cdot \frac{x_i}{\bar{x}} - C_i x_i \qquad \text{if } \bar{x} > 0$$
$$= 0 \qquad \qquad \text{if } \bar{x} = 0$$

a) Prove that every strategy that is large enough is dominated for Player i. b) Fixing the set $I \subset \{1, \ldots, n\}$ of active players (those with $x_i > 0$) write the system characterizing a possible Nash equilibrium. Then give the conditions on the parameters which guarantee the feasibility of that system.

c) Study the local stability of the equilibria found in b).

4) Stability by sequential Cournot tatonnement

Given an ordering $\{1, 2, \ldots, n\}$ of the players, the $\{1, \ldots, n\}$ sequential Cournot tatonnement starting at x^o is the sequence $x^o, x^1, \ldots, x^t, \ldots$ where

$$x_i^t = r_i(x_1^t, \ldots, x_{i-1}^t, x_{i+1}^{t-1}, \ldots, x_n^{t-1})$$

if $t = i$ modulo n and $x_i^t = x_i^{t-1}$ otherwise.

Say that a Nash equilibrium x^* is $\{1, \ldots, n\}$ stable if for any initial position x^o the $\{1, \ldots, n\}$ sequential Cournot tatonnement starting at x^o converges to x^*. a) If $n = 2$, the $\{1, 2\}$ stability, the $\{2, 1\}$ stability, and the stability in the sense of Definition 1 all coincide. b) If $n \geq 3$, these notions differ. Consider, for instance, the three players game

$$X_i = R, \ i = 1, 2, 3 \qquad \begin{aligned} u_1(x) &= -(x_1 - x_2)^2 \\ u_2(x) &= -(x_2 - \tfrac{1}{3} x_3)^2 \\ u_3(x) &= -(4x_1 - 3x_2 - x_3)^2 \end{aligned}$$

The unique Nash equilibrium is $(0, 0, 0)$. Prove that it is not $\{1, 2, 3\}$ stable, whereas it is $\{2, 1, 3\}$ stable. Is it stable in the sense of Definition 1?

5) <u>Cournot tatonnement and sophisticated equilibrium</u> (Moulin [1984])

a) In the Cournot duopoly with constant return to scales, show that the Nash equilibrium is the sophisticated equilibrium as well. b) Let G be a two-person game where $X_i = [0, 1]$, u_i is continuous on $[0, 1]^2$ and <u>strictly</u> concave w. r. t. x_i, $i = 1, 2$. Thus the best reply of each player is single valued and continuous. Show that for any subintervals $Y_i \subset [0, 1]$, $i = 1, 2$ we have

$$\mathcal{D}_i(u_i, Y) = \underset{Y_i}{\text{proj}} \ r_i(Y_j)$$

Deduce that the successive elimination of dominated strategies is given by

$$X^{t+1}_i = r_i(X^t_j) \qquad\qquad t = 1, 2, \ldots$$

Conclude that G is cominance solvable iff it has a unique (globally), stable Nash equilibrium.

CHAPTER 7. MIXED STRATEGIES

1. MIXED EXTENSION OF A GAME

In some normal-form games, although each player can pick
at will any element from the strategy set that has been
given, each should incorporate a rational tactical move
involving a voluntary randomness within these choices.

Example 1.   de Montmort's game.

In the late eighteenth century, the French mathematician
René de Montmort described the following situation. To make a
gift to his son, a father decided "I shall have a gold coin
in my right hand or my left hand and you shall name a hand;
if your guess is correct and it is my right hand you shall

get the gold coin; if your guess is correct and it is my
left hand, you shall receive two gold coins; otherwise you
shall get nothing." Then Montmort asks how much this gift
is actually worth to the son while pointing out that "in
this game if the players are equally shrewd and clear-sighted,
there is no way we can lay down a rule of conduct." That
is, no optimal strategy exists in this game. In fact, a
contemporary of Montmort, James Waldegrave, did propose to
use mixed strategies in a similar game. It is not clear,
however, that he had the intuition of optimal mixed strategies
(see the historical article by Rives [1975]).

The following 2 x 2, two-person, zero-sum game is in order.

	Left	Right
Left	2	0
Right	0	1

Son (rows), Father (columns)

(Numbers measure the son's payoffs.) The game has no value
$\alpha_1 = 0 < \alpha_2 = 1$ and our players compete for the second move.
The natural way to prevent one's own choice from being outguessed
is to make this choice at random. That is, instead of choosing

a strategy x_1, such as left for sure or right for sure, the son will use a randomized strategy μ_1 taking the value left or right with respective probability p_1, $1 - p_1$. Imagine he designs a coin that shows L with probability p_1 and tosses this coin to choose his actual strategy. Assume that the father can never observe the actual draw of the coin. This implies that the father's strategy, whether or not it is a random variable, cannot be correlated to the son's strategy. Therefore the expected payoff to the son of the so-called "mixed strategy" μ_1 is at least inf $\{2 p_1, 1 - p_1\}$. By choosing $p_1 = 1/3$, the son is guaranteed of an expected payoff at least two thirds.

The situation of the father is symmetrical. By playing at random strategies left and right with respective probabilities p_2, $1 - p_2$, his maximal expected loss is sup $\{2 p_2, 1 - p_2\}$.

Choosing $p_2 = 1/3$, the father is guaranteed of an expected loss <u>at most</u> of two thirds. Thus, incorporating tactical uncertainty into strategical choices suggests the value of two third as a fair estimate of the father's generosity.

<u>Definition 1.</u>

Let $G = \{X_i, u_i; i = 1, \ldots, n\}$ be an N-normal form game where X_i is a finite set for all $i \in N$. A <u>mixed strategy</u> of Player i is a probability distribution μ_i over X_i. Hence, the set M_i of Player i's mixed

strategies is the unit simplex of R^{X_i}. *The* mixed
extension of G is the N-normal-form game G_m =
$(M_i, \bar{u}_i; i = 1, \ldots, n)$ *where for all* $\mu \in M_N$

$$\bar{u}_i(\mu) = \sum_{x \in X_N} u_i(x)\mu_1(x_1) \cdot \mu_2(x_2) \cdot \ldots \cdot \mu_n(x_n) \qquad (1)$$

In the mixed extension G_m of G, Player i's strategy is a
probability distribution μ_i over X_i. It is understood that
Player i actually constructs a private lottery that selects
a strategy $x_i \in X_i$ according to μ_i. Privacy of the lottery means
that Player i alone is informed of the strategy x_i actually
selected by the lottery (even if the other players may know
the probability distribution μ_i). Moreover, i's lottery is
stochastically independent from j's lottery, for all j, j $\neq$ i
(so that j can deduce no information on i's lottery by
observing his own lottery). As the random variables are
mutually independent, $\bar{u}_i(\mu)$ is the expected utility of Player i.
Taking $\bar{u}_i$ as the payoff function of i in G_m amounts to saying
that Player i compares various lotteries μ, μ' on X_N by simply
comparing their associated expected utility $\bar{u}_i(\mu)$, $\bar{u}_i(\mu')$.
In other words, u_i is a von Neumann-Morgenstern utility function
that summarizes the preferences of Agent i over all conceivable lotteries on X_N. More justifications of this
assumption pertain to the statistical theory of decision and
will not be pursued here. See Luce and Raiffa [1957].

Notice that the cardinality of $u_i(x)$ plays a fundamental role here, whereas in all previous chapters, only the ordinal preference orderings induced by the u_i's over X_N did matter.

A pure strategy $x_i \, \varepsilon \, X_i$ of Player i in the initial game G will be identified with the mixed strategy $\delta_{x_i} \, \varepsilon \, M_i$ which selects x_i with probability one:

$$\delta_{x_i}(x'_i) = 0 \qquad \text{all } x'_i \, \varepsilon \, X_i, \; x'_i \neq x_i$$

$$\delta_{x_i}(x_i) = 1$$

Namely, formula (1) implies at once

$$\bar{u}_i(\delta_x) = u_i(x) \qquad \text{for all } x \, \varepsilon \, X_N \text{ and}$$

$$\delta_x = (\delta_{x_i})_{i \, \varepsilon \, N}$$

Thus we look at X_i as a subset of M_i and at $\bar{u}_i$ as the extension of u_i from X_N to M_N.

Theorem 1.

If X_i is finite for all $i \, \varepsilon \, N$, the set of Nash equilibrium outcomes of G_m is a nonempty compact subset of M_N. Moreover, it contains the set of Nash equilibrium outcomes of G:

$$NE(G) \subset NE(G_m) \neq \emptyset$$

Proof.

Pick any NE outcome x of G. Denoting

$$\delta_{x_{-i}} = (\delta_{x_j})_{j \neq i} \quad \varepsilon \quad M_{N \setminus \{i\}}$$

we observe that

$$\sup_{\mu_i \varepsilon M_i} \overline{u}_i(\mu_i, \delta_{x_{-i}}) = \sup_{y_i \varepsilon X_i} u_i(y_i, x_{-i})$$

M_i is the convex hull of X_i (when the latter is identified with a subset of M_i), and $\overline{u}_i$ is linear with respect to its variable μ_i. Now, by the NE property of x,

$$\sup_{y_i \varepsilon X_i} u_i(y_i, x_{-i}) = u_i(x) = \overline{u}_i(\delta_{x_i}, \delta_{x_{-i}})$$

Combining these properties yields that δ_x is a NE outcome of G_m.

Check next that G_m satisfies all the assumptions of Nash's theorem (Theorem 2, Chapter 5). That maintained that M_i is convex and compact within R^{X_i}, $\overline{u}_i$ is continuous over M_N and linear with respect to its variable μ_i. QED.

Just as in the general case, we have also Nash's theorem for symmetrical mixed games.

Nash's theorem for symmetrical mixed games.

A symmetrical, finite game has at least one symmetrical Nash equilibrium in mixed strategies.

The next two sections propose computational methods for finding the mixed equilibria in two-person, zero-sum games (Section 2) or in general n-person games (Section 3).

154

2. FINITE TWO-PERSON, ZERO-SUM GAMES

Finite two-person, zero-sum games are converted into inessential games by resorting to mixed strategies. Here the mixed extension is very helpful.

Corollary to Theorem 1.

If $G = (X_1, X_2, u_1, -u_1)$ is a finite two-person, zero-sum game, its mixed extension is a two-person, zero-sum game G_m with a value $v_m(u_1)$ called the mixed value of G. We have

$$\sup_{x_1} \inf_{x_2} u_1 \leq v_m(u_1) \leq \inf_{x_2} \sup_{x_1} u_1 \qquad (2)$$

Also, in G_m, Player i's optimal strategies form a (nonempty) compact convex subset of M_i.

Proof.

By Theorem 1, the two-person, zero-sum game has at least one Nash equilibrium; therefore it has a value $v_m(u_1)$ (see the end of Section 3, Chapter 5).

Consider a pure strategy $x_1 \in X_1$ and a mixed strategy $\mu_2 \in M_2$. Since $\bar{u}_1(\delta_{x_1}, \mu_2)$ is the expected value of $u_1(x_1, .)$ w. r. t. μ_2, we have

$$\inf_{x_2 \in X_2} u_1(x_1, x_2) \leq \bar{u}_1(\delta_{x_1}, \mu_2)$$

Since μ_2 is arbitrary, this implies

$$\inf_{x_2} u_1(x_1, x_2) \leq \inf_{\mu_2} \bar{u}_1(\delta_{x_1}, \mu_2) \leq \sup_{\mu_1} \inf_{\mu_2} \bar{u}_1(\mu_1, \mu_2)$$

Since x_1 is arbitrary, we get

$$\sup_{x_1} \inf_{x_2} u_1 \leq \sup_{\mu_1} \inf_{\mu_2} \bar{u}_1 = v_m(u_1)$$

A symmetrical argument yields the right-hand inequality in (2).

To prove the last statement, we use the fact that $\mu_1^* \in M_1$ is optimal for Player 1 in G_m iff it is prudent (Chapter 1, Theorem 1) namely iff it satisfies

$$\phi(\mu_1) = \inf_{\mu_2} \bar{u}_1(\mu_1, \mu_2) \geq v_m(u_1)$$

Since $\bar{u}_1(\cdot, \mu_2)$ is a concave function (in its variable μ_1), so is the function ϕ (the infimum of concave functions is concave as well). Hence any set such as $\{\mu_1 \in M_1 / \phi(\mu_1) \geq \lambda\}$ must be convex. Since $\bar{u}_1(\cdot, \mu_2)$ is a continuous function, then ϕ is upper semicontinuous. Thus, it reaches its maximum over a compact subset of M_1. QED.

Remark 1.

The argument used to prove the corollary to Theorem 1, above, can be used to prove that, under the assumptions of von Neumann's theorem (Theorem 2, Chapter 1), the set of optimal strategies of either player is convex.

If the initial game G has a value, its mixed extension has the same value (and optimal strategies in G_m contain all convex combinations of optimal strategies in G). If, in contrast, G has no value, its mixed value $v_m(u_1)$ lies within the duality gap $\sup\inf u_1$, $\inf\sup u_1$. In general, both inequalities in (2) are strict (see Exercise 11).

Example 2. 2 x 2 games.

Consider the game

$$\text{Player 1}\quad \begin{matrix}\text{Top}\\[20pt]\text{Bottom}\end{matrix}\begin{bmatrix} a & c \\ & \\ b & d \end{bmatrix}$$

$$\begin{matrix}\text{Left} & \text{Right}\end{matrix}$$

$$\text{Player 2}$$

It has no value iff [a, d] ∩ [b, c] is empty (as an exercise, prove this claim). In this latter case each player has a unique optimal strategy, namely,

$$\text{set } \Delta = a + d - b - c$$

$$\mu^*_1 = (\frac{d-b}{\Delta}, \frac{a-c}{\Delta}) \quad \mu^*_2 = (\frac{d-c}{\Delta}, \frac{a-b}{\Delta})$$

$$\quad\quad\text{Top Bottom}\quad\quad\text{Left Right}$$

The mixed value is worth

$$v_m = \frac{ad - bc}{\Delta} = \frac{ad - bc}{a + d - b - c}$$

These facts are a consequence of Lemma 1 below.

Here are a few remarks that help computing the mixed value and optimal mixed strategies.

a) <u>2 x p games</u>

Player 1 has two pure strategies. His optimal mixed strategy is easily computed

$$u_1 = \begin{bmatrix} a_1 & \cdots & a_p \\ b_1 & \cdots & b_p \end{bmatrix}$$

$$v_m(u_1) = \sup_{0 \le \lambda \le 1} \left\{ \inf_{1 \le k \le p} (\lambda a_k + (1 - \lambda) b_k) \right\} = \sup_{0 \le \lambda \le 1} \phi(\lambda)$$

where ϕ is a concave function of which the graph is easily drawn. For instance

$$\begin{bmatrix} 0 & 2 & 5 & 3 \\ 6 & 1 & 0 & 2 \end{bmatrix}$$

has a mixed value $12/7$, the optimal strategy of Player 1 $\mu_1 = 5/7$ top $+ 2/7$ bottom. Hence, μ_2 (being a best reply to μ_1) weighs only the two left strategies

$$\mu_2 = (1/7, 6/7, 0, 0)$$

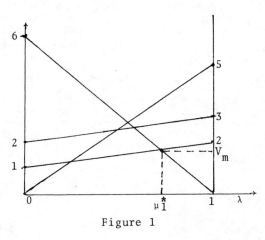

Figure 1

The graph of ϕ

b) Eliminating dominated strategies

    If a player has a strictly dominated (pure) strategy x_i (for some y_i we have $u_i(x_i, x_j) < u_i(y_i, x_j)$ all x_j), then any one of his optimal strategies gives zero probability to that strategy. If a player has a dominated strategy, he has at least one optimal strategy that gives zero probability to that strategy. In both cases we can drop the dominated strategy and compute the value of the reduced game. Here is an example.

$$\begin{bmatrix} 1 & 2 & 3 & 4 \\ 3/2 & 0 & 1 & 0 \\ 0 & 1 & 4 & 3 \end{bmatrix}$$

which reduces to

$$\begin{bmatrix} 1 & 2 \\ 3/2 & 0 \end{bmatrix}$$

The value is 6/5 (see Example 1).

c) Completely mixed equilibrium

The carrier of a mixed strategy μ_i is the set of pure strategies x_i in X_i such that $\mu_i(x_i) > 0$. It is denoted $[\mu_i]$. A mixed strategy is completely mixed if its carrier has all pure strategies in it: $[\mu_i] = X_i$.

Lemma 1.

Let $G = (X_1, X_2, u_1)$ be a game and (μ^*_1, μ^*_2) be a pair of mixed strategies satisfying

$$\bar{u}_1(\mu^*_1, \mu_{x_2}) = \bar{u}_1(\mu^*_1, \delta_{y_2}) \qquad all \ x_2, y_2 \ \varepsilon \ X_2$$

(3)

$$\bar{u}_1(\delta_{x_1}, \mu^*_2) = \bar{u}_1(\delta_{y_1}, \mu^*_2) \qquad all \ x_1, y_1 \ \varepsilon \ X_1$$

Then (μ^*_1, μ^*_2) is a saddle point of G_m, and these numbers all equal the mixed value. Suppose, moreover, that the payoff matrix of the game $U = [u_1(x_1, x_2)]_{\substack{x_1 \ \varepsilon \ X_1 \\ x_2 \ \varepsilon \ X_2}}$ is nonsingular. Then the mixed value is worth

$$v_m(u_1) = \frac{det \ U}{\sum_{(x_1, x_2) \ \varepsilon \ X_1 \ x \ X_2} \tilde{u}(x_1, x_2)}$$

(4)

where $\tilde{u}(x_1, x_2)$ is the cofactor of the entry (x_1, x_2) in U. Conversely if (μ_1, μ_2) is a saddle point of completely mixed strategies, then it satisfies system (3).

Proof.

Suppose $(\mu^*_1, \mu^*_2) \in M_1 \times M_2$ satisfies system (3) and denote $\alpha_1 = \bar{u}_1(\mu^*_1, \delta_{x_2})$ all $x_2 \in X_2$, and $\alpha_2 = \bar{u}_1(\delta_{x_1}, \mu^*_2)$ all $x_1 \in X_1$. By the linearity of $\bar{u}_1$ in its second variable (respectively the first variable), this implies $\alpha_1 = \bar{u}_1(\mu^*_1, \mu_2)$ all $\mu_2 \in M_2$ (respectively $\alpha_2 = \bar{u}_1(\mu_1, \mu^*_2)$ all $\mu_1 \in M_1$). Therefore (μ^*_1, μ^*_2) is a saddle point of G_m, and $\alpha_1 = \alpha_2 = v_m(G)$.

We prove now formula (4). In matrix notation, the lower equations in (3) are

$$U \mu_2 = v_m \cdot \mathbf{1}$$

where μ_2 is the column vector $[\mu_2(x_2)]_{x_2 \in X_2}$ and $\mathbf{1}$ is the column vector with all its coordinates equal to 1. Since U is nonsingular, this amounts to

$$\mu_2 = v_m U^{-1} \cdot \mathbf{1}$$

As μ_2 is a mixed strategy, $^t\mathbf{1} \cdot \mu_2 = \sum_{x_2 \in X_2} \mu_2(x_2) = 1$. Therefore $1 = v_m \cdot {}^t\mathbf{1}U^{-1}\mathbf{1}$ or equivalently

$$v_m = \frac{1}{{}^t\mathbf{1}U^{-1}\mathbf{1}}$$

The comatrix $\tilde{U}$ is related to the inverse matrix by

$U^{-1} = (1/\det U) \cdot {}^t\tilde{U}$, so by substituting $\tilde{U}$ for U^{-1} we have proved the lemma.

Conversely, suppose (μ_1^*, μ_2^*) is a completely mixed saddle point of $\bar{u}$. This implies, in particular,

$$\bar{u}(\mu_1^*, \mu_2^*) = \sum_{x_1 \varepsilon X_1} \mu_1^*(x_1) \cdot \bar{u}(\delta_{x_1}, \mu_2^*) \geq \bar{u}(\delta_{y_1}, \mu_2^*)$$

$$\text{for all } y_1 \varepsilon X_1$$

Multiplying each inequality by $\mu_1^*(y_1)$ and summing over y_1 gives an equality. Since all $\mu_1^*(y_1)$ are strictly positive, this means that all inequalities actually are equalities and the lower half of system (3) is proved. The upper half is proved symmetrically. QED.

System (3) allows us to compute any saddle point in mixed strategies as soon as its carrier $[\mu_1] \times [\mu_2]$ is known. In the restriction of our game to $[\mu_1] \times [\mu_2]$, the pair (μ_1, μ_2) is but a completely mixed saddle point. System (3) does not, however, allow for an independent computation of one player's optimal strategies (see Exercise 8).

Notice that system (3) has $|X_2| - 1 + |X_1| - 1$ equations to determine the same number of unknown variables. Yet, if $|X_2| < |X_1|$, it gives $|X_1| - 1$ equations to find $|X_2| - 1$ unknown (i.e., mixed strategy μ_1). Hence, in general, a completely mixed saddle point will be found only for games where $|X_1| = |X_2|$ (a rigorous genericity argument can be made by proving that this statement holds true on an open, dense subset of pairs of payoff matrices). We give three more examples.

Example 3.

0	3	2
3	0	2
1	1	0

There is no completely mixed saddle point. The carrier of the unique saddle point is the northwest 2 x 2 game. The value is 3/2 and optimal strategies are (1/2, 1/2, 0) (see Example 1).

Example 4. la Bataille.

Each player has the same hand {0, 1, 2, ..., 9}. Each round consists of throwing away one card. The player with the highest card is paid $1 by the other (no payments are made if they show the same number). The final payoff is the sum of these ten payoffs. Because the (mixed) game is symmetrical, its value must be zero (see Exercise 9, Chapter 1) and both players must have the same optimal strategies. The question is: How to play optimally? For instance, suppose after four rounds the remaining hands are

Player 1's hand H_1 = {2, 4, 5, 7, 8, 9}
Player 2's hand H_2 = {0, 1, 3, 4, 6, 7}

Since each player knows his opponent's hand, Player 1 may have a way to exploit his comparatively stronger hand. In fact, he does not, since the strategy of drawing one card at random with uniform probability is optimal for both players. To see this, describe a pure strategy of Player i after four rounds as a one-to-one mapping x_i from $\{1, 2, \ldots, 6\}$ into H_i ($x_2(1) = 4$ means Player 2 throws away card 4 first). Assuming Player 2 picks at random one such mapping with uniform probability (denoted as a mixed strategy μ_2) we prove that all pure strategies of Player 1 are equivalent. By Lemma 1, this will establish the claim.

Compute

$$\bar{u}_1(x_1, \mu_2) = \frac{1}{6!} \sum_{x_2} u_1(x_1, x_2) = \frac{1}{6!} \sum_{x_2} \sum_{1 \leq k \leq 6} \operatorname{sgn}(x_1(k) - x_2(k))$$

where $\operatorname{sgn}(z)$ is $+1$ if $z > 0$, -1 if $z < 0$, and 0 if $z = 0$. Thus

$$\bar{u}_1(x_1, \mu_2) = \frac{1}{6!} \sum_{1 \leq k \leq 6} [\#\{x_2 | x_2(k) < x_1(k)\} - \#\{x_2 | x_2(k) > x_1(k)\}]$$

$$\bar{u}_1(x_1, \mu_2) = \sum_{a \in H_1} \frac{1}{6} [\#\{b \in H_2 | b < a\} - \#\{b \in H_2 | b > a\}]$$

where the latter expression does not depend on x_1.

Example 5. Cheater-inspector.

The cheater-inspector game is an example of a recursive game where the mixed value and optimal strategies are computed inductively. There are n rounds. The cheater can cheat only

once, and the inspector can inspect in every round. If the
cheater cheats without being inspected, the game stops and
he gains $1 in every remaining period (e.g., successfully
cheating the seventh round of ten gains him $4). If the
cheater is inspected while cheating, the game stops, and he
is fined $a at every remaining period. If he does not cheat,
he receives $b from the insepctor if inspected and nothing
if not inspected. In both cases the game goes on to the next
round.

Let G_n be the n-round game with value v_n. The recursive
formula is

$$G_n = \begin{array}{|c|c|} \hline n & -na \\ \hline G_{n-1} & b+G_{n-1} \\ \hline \end{array}$$

cheats

does not cheat

does not inspect | inspects

with G_0 = the constant zero-payoff game. Hence, v_n is
computed inductively

$$v_0 = 0; \quad v_n = \text{value} \left\{ \begin{array}{|c|c|} \hline n & -na \\ \hline v_{n-1} & b+v_{n-1} \\ \hline \end{array} \right\}$$

Clearly $0 \leq v_{n-1} \leq n$. Therefore

$$v_n = \frac{n(b + v_{n-1}) + anv_{n-1}}{n + b + an} \Rightarrow v_n = n \frac{b}{1 + b + a}$$

Thus the average value is constant in n. Also the optimal strategy is to cheat in round n with probability (b)/(b + n(1 + a)) and to inspect with probability

$$\frac{n(a + b)(1 + a)}{[n(1 + a) + b][1 + a + b]}$$

3. FINITE N-PLAYER GAMES

We have seen that a game in pure strategies might have a unique Nash equilibrium where each player is using a dominated strategy (Example 1, Chapter 5). Fortunately this situation cannot arise in mixed games.

Lemma 2.

In the mixed extension of the finite game $G = (X_i, u_i, i = 1, \ldots, n)$ there is a Nash equilibrium $\mu = (\mu_i, i = 1, \ldots, n)$ where no player is using a dominated pure strategy with positive probability

$$\mu_i(x_i) > 0 \Rightarrow x_i \in D_i(u_i, X) \quad all\ i \in N,\ x_i \in X_i$$

This statement can be strengthened by requiring that any x_i in the carrier of μ_i is undominated in the mixed extension of G:

$$\mu_i(x_i) > 0 \Rightarrow \delta_{x_i} \in D_i(\overline{u}_i, M)$$

Proof (We give only the sketch of the tedious proof).

Observe first that $D_i(\overline{u}_i, M) = N_i$ is nonempty (since M is compact and $\overline{u}_i$ are continuous). Next take $\mu_i \in N_i$ and check that any δ_{x_i} such that x_i is in the carrier of $\mu_i(\mu_i(x_i) > 0)$ must be in N_i as well (as an exercise, find out why.) Thus, $Y_i = \{x_i \in X_i / \delta_{x_i} \in N_i\}$ is a nonempty subset of X_i. Restrict the initial game to $Y_1 \times \ldots \times Y_n$ and pick a mixed-Nash equilibrium of that game, say μ. Then μ is a Nash equilibrium of G_m as well, since any mixed strategy μ'_i such that $[\mu'_i] \neq Y_i$ must be dominated by some μ''_i such that $[\mu''_i] \subset Y_i$. QED.

Another good feature of mixed-Nash equilibria is individual rationality, as is demonstrated in Lemma 3.

Lemma 3.

Let G_m be the mixed extension of the finite game $G = (X_i, u_i, i = 1, \ldots, n)$. At a Nash equilibrium μ^* of G_m, each player enjoys at least his mixed, secure utility level, that is, the mixed value of the two-person, zero-sum game (X_i, X_{-i}, u_i). For all $\mu^* \in NE(G_m)$, all $i = 1, \ldots, n$.

$$v_m(u_i) = \inf_{\mu_{-i}} \sup_{\mu_i} u_i(\mu) \leq u_i(\mu^*)$$

We omit the straightforward proof.

Unfortunately, it is often the case that a mixed-Nash equilibrium i) gives to each player no more than his mixed secure utility level, ii) does so without even

guaranteeing that level to each agent, since the Nash
equilibrium strategy is not prudent. Here is an example.

Example 6. The crossing game (continued).

Recall the matrix of this 2 x 2 game introduced as Example 1,
Chapter 2:

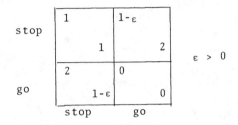

There are two Nash equilibria in pure strategies, namely
(stop, go) and (go, stop). Since the game is symmetrical,
there must be a symmetrical Nash equilibrium in mixed strategies.
It is a completely mixed Nash equilibrium, namely,

$$\mu^*_1 = \mu^*_2 = \frac{1 - \varepsilon}{2 - \varepsilon} \text{ stop} + \frac{1}{2 - \varepsilon} \text{ go}$$

with corresponding utility level $(2 - 2\varepsilon)/(2 - \varepsilon)$ for each
player. Notice that each player "goes" with a probability
slightly above 1/2 (hence an accident occurs with probability
$\geq 1/4$ and each get hardly any better a utility level than his
secure utility level, namely, $1 - \varepsilon$. In fact, μ^*_1 guarantees
only $(1 - \varepsilon)^2/(2 - \varepsilon) \approx 1/2$ to each player (if the opponent
decides to go no matter what). Hence the price for a tiny
increase in utility is to incur a high risk.

The point is even more striking when we look at the case where each player prefers that the other goes, given that he stops.

	stop	go
stop	1 1	1+ε' 2
go	2 1+ε'	0 0

$\varepsilon' > 0$

Now the guaranteed utility level is $(2 + 2\varepsilon')/(2 + \varepsilon')$. Actually this is <u>the same</u> as the utility at the completely mixed Nash equilibrium of $\mu^*_1 = \mu^*_2 = (1 + \varepsilon')/(2 + \varepsilon')$ stop + $1/(2 + \varepsilon')$ go. Again, μ^*_1 is more risky than the prudent strategy $\tilde{\mu}_1$ ($\tilde{\mu}_1 = (2/(2 + \varepsilon'))$ stop + $(\varepsilon'/(2 + \varepsilon'))$ go), but no more profitable either.

$$\frac{(1 + \varepsilon')^2}{2 + \varepsilon'} = \inf_{\mu_2} u_1(\mu^*_1, \mu_2) < u_1(\mu^*_1, \mu^*_2) = u_1(\tilde{\mu}_1, \mu^*_2)$$

$$= \inf_{\mu_2} u_1(\tilde{\mu}_1, \mu_2) = \frac{2 + 2\varepsilon'}{2 + \varepsilon'}$$

Here the tradeoff is between a stable outcome (the mixed Nash equilibrium) that involves genuine risk if your opponent does not stick to it, and a prudent outcome (the pair of prudent mixed strategies) where no one is taking any unnecessary risk but every one gives to his opponent an incentive to deviate. Exercise 12 gives another instance of this conflict.

We state now the parallel result of Lemma 1 for n players games. Recall that $[\mu_i]$ denotes the carrier of a mixed strategy $\mu_i \in M_i$.

Lemma 4.

Let the initial game $G = (X_i, u_i; i \in N)$ be given with finite strategy sets and let $\mu^* \in NE(G_m)$ be a mixed Nash equilibrium. Then the following system holds true.

$$\forall i \in N \forall x_i \in [\mu^*_i] \; u_i(\delta_{x_i}, \mu^*_{-i}) = u_i(\mu^*) \tag{5}$$

We do not repeat the identical proof. For two-person games, formula (4) easily can be generalized (see exercise 14).

The point of system (5) is this. At a mixed Nash equilibrium μ Player i's best replies to μ_{-i} contains all mixed strategies μ'_i with the same carrier as μ_i. In particular, if μ is a completely mixed equilibrium, then any strategy of Player i is a best reply to the $N\setminus\{i\}$-tuple μ_{-i} of equilibrium strategies by the other players. In particular at a completely mixed equilibrium of a two-person game, a Player i picks a strategy that makes Player j indifferent about choosing any one from all of his (mixed) strategies, quite independently of his own payoff u_i. This is illustrated by Example 6 above. However, in a game with at least three players the completely mixed equilibrium

strategies must be determined jointly, so that μ_i depends on
the payoff function u_i through system (5).

When the sets $[\mu_i] = Y_i$ are given, system (5) provides
$\sum_{i \in N} |Y_i| - 1$ independent equations for determining $\mu_1, \ldots, \mu_n$,
with the same number of unknown variables. Thus, in general,
system (5) has at most one solution. Accordingly, a finite
game G has in general, finitely many mixed Nash equilibria.

The case of two-person games

In two-person games, one proves that for an open dense
subset of games i) The number of mixed Nash equilibria
(including those in pure strategies) is <u>odd</u> (this is a
technically difficult result; see Lemke and Howson [1964]).
ii) For each Nash equilibrium (μ_1, μ_2) the carriers $[\mu_1]$ and
$[\mu_2]$ have the same size. iii) The mixed Nash equilibria
($|[\mu_i]| \geq 2$) are Pareto-inferior outcomes of the mixed game.

For instance, consider a 2 x 2 game such as

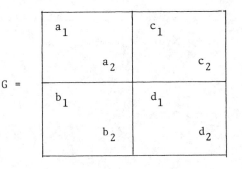

where a_i, b_i, c_i, d_i are four pairwise distinct numbers
($i = 1, 2$). Then exactly one of the following two cases must
arise. a) G has one Nash equilibrium (in pure strategies)
and no other mixed Nash equilibria. b) G has zero or two
Nash equilibria in pure strategies and one completely mixed
Nash equilibrium μ^* given by the formulas

$$\mu^*_1 = \left(\frac{d_2 - b_2}{\Delta_2}, \frac{a_2 - c_2}{\Delta_2} \right)$$

$$\Delta_i = a_i + d_i - b_i - c_i$$

$$\mu^*_2 = \left(\frac{d_1 - c_1}{\Delta_1}, \frac{a_1 - b_1}{\Delta_1} \right)$$

with corresponding utility level $u^*_i = (a_i d_i - b_i c_i)/\Delta_i$.

4. INFINITE GAMES

This section is significantly more technical than the
rest of Chapter 7. The discussion involves games where the
initial strategy sets are infinite. Mixed-strategy equilibria are much more difficult to compute. In our first
example, the symmetrical nature of the game makes this
computation tractable.

Example 7.   The mixed-auction dollar game.

In the game of Example 2, Chapter 2, the players select
their bids according to independent probability distributions.

We recall the normal form of the game: A pure strategy is a bid $x_i \geq 0$; thus $X_i = [0, +\infty[$. For any outcome x, denote $x(i) = \sup_{j \neq i} x_j$. The payoff functions $u_1, \ldots, u_n$ are such that

$$u_i(x) = 1 - x(i) \qquad \text{if } x_i > x(i)$$

$$= -x_i \qquad \text{if } x_i < x(i)$$

Note that we do not specify the tie-breaking rule when several bids are highest. Actually this event has a probability of zero at the mixed equilibrium.

We look for a symmetrical, completely mixed equilibrium. That is to say, we seek a probability distribution μ^* on $[0, +\infty[$ with density f in the Lebesgue measure, such that

$$\bar{u}_1(\mu_1, \mu^*, \ldots, \mu^*) \leq \bar{u}_1(\mu^*, \mu^*, \ldots, \mu^*)$$

for all probability distributions μ_1. As in Lemma 4 above, this amounts to

$$\phi(x_1) = \bar{u}_1(\delta_{x_1}, \underbrace{\mu^*, \ldots, \mu^*}_{n-1}) \text{ does not depend on } x_1 \qquad (6)$$

Let F denote the cumulative distribution of μ^* ($F(x) = \mu^*([0, x])$). When players $2, \ldots, n$ use independently μ^*, the cumulative distribution of $x(1)$ is F^{n-1}. Therefore,

$$\phi(x_1) = E_{[0,x_1]}(1 - x(1)) - x_1 \operatorname{Prob}(x(1) \geq x_1)$$

$$= \int_0^{x_1} (1 - t) dF^{n-1}(t) - x_1(1 - F^{n-1}(x_1))$$

Condition (6) means $\phi^1(x_1) = 0$ all x_1, namely

$$0 = (1 - x_1)(F^{n-1})'(x_1) - \{1 - F^{n-1}(x_1) - x_1(F^{n-1})'(x_1)\}$$

$$<=> (F^{n-1})'(x_1) + F^{n-1}(x_1) = 1$$

Together with $F(0) = 0$ this gives $F(x) = (1 - e^{-x})^{1/n - 1}$.
Hence the equilibrium density

$$f(x) = \frac{1}{n-1} \frac{e^{-x}}{(1 - e^{-x})^{(n-1)/(n-2)}}$$

Note that for any $\lambda > 0$ the probability that the winning bid
exceeds λ is nonzero. However, the expected winning bid is
finite. Also, the equilibrium payoff is zero for each
player (since $\phi(0) = 0$). Thus, using the equilibrium strategy
does no more than making the prudent zero bid (just as in
Example 6 above). Exercise 21 proposes the first price variant
of the game discussed above.

Our next two examples show that, in infinite games, mixed
strategies may fail to ensure the existence of a Nash
equilibrium. Both examples are two-person, zero-sum games.

Example 8.   Chinese poker.

In Chinese poker each of the two players picks a (nonnegative)
integer. The player who called the largest number wins a
dollar:

$$
X_1 = X_2 = \mathbb{N} \quad u_1(x_1, x_2) = \begin{cases} +1 & \text{if } x_2 < x_1 \\ 0 & \text{if } x_2 = x_1 \\ -1 & \text{if } x_1 < x_2 \end{cases}
$$

A mixed strategy is a probability distribution μ_i on $\mathbb{N}$. The mixed payoff is defined as usual. In fact using mixed strategies does not reduce the duality gap <u>at all</u>.

$$
\sup_{\mu_1} \inf_{\mu_2} \bar{u}_1 = -1 < +1 = \inf_{\mu_2} \sup_{\mu_1} \bar{u}_1 \tag{7}
$$

To check this, fix a probability distribution μ_1 on $\mathbb{N}$:

$$
\mu_1(n) \geq 0 \qquad \text{for all } n = 1, 2 \ldots
$$

<u>and</u>

$$
\sum_{n=1}^{+\infty} \mu_1(n) = 1
$$

For any $\varepsilon > 0$, there is an integer n_ε such that $\sum_{n \geq n_\varepsilon} \mu_1(n) \leq \varepsilon$. Then compute

$$
\bar{u}_1(\mu_1, \delta_{n_\varepsilon}) = - \sum_{n < n_\varepsilon} \mu_1(n) + \sum_{n > n_\varepsilon} \mu_1(n) < -1 + 2\varepsilon
$$

Since ε is arbitrary this implies $\inf_{\mu_2} u_1(\mu_1, \mu_2) \leq -1$. The left-hand equality in (7) follows because μ_1 is arbitrary as well.

In Example 8 both the pure-strategy sets and the mixed-strategy sets are not compact, which is the reason why von

Neumann's theorem does not apply (even though all other
assumptions hold true). In our next example all strategy
sets, pure and mixed, are compact, but the payoff function
is not continuous.

Example 9. A fund-allocation problem (Sion and Wolfe [1957]).

Each player must allocate his unit campaign budget between
two states A, B. In A the player with the largest budget wins.
In B Player 2 (the incumbent) has a premium which is worth 1/2
(thus Player 1 wins B only if $y_1 > y_2 + 1/2$ where y_i is the
budget allocated to B by Player i). The challenger needs only
to win in one state to win the overall game. There is a draw
if he matches the incumbent's budget in one state. Hence the
normal-form game $X_1 = X_2 = [0, 1]$, x_i is the amount spent by
Player i in A (thus $y_i = 1 - x_i$).

$$
\begin{aligned}
u_1(x_1, x_2) &= -1 && \text{if } x_1 < x_2 < x_1 + \tfrac{1}{2} \\
&= 0 && \text{if } x_1 = x_2 \text{ or } x_2 = x_1 + \tfrac{1}{2} \\
&= +1 && \text{if } x_2 < x_1 \text{ or } x_1 + \tfrac{1}{2} < x_2
\end{aligned}
$$

In the mixed extension of this game, (where each uses a Radon
probability measure on [0, 1]) we claim that

$$
\sup_{\mu_1} \inf_{\mu_2} \bar{u}_1 = \frac{1}{3} < \frac{3}{7} = \inf_{\mu_2} \sup_{\mu_1} \bar{u}_1
$$

Suppose, first, μ_2 is a probability distribution on X_2

such that $\sup\limits_{x_1} u_1(\delta_{x_1}, \mu_2) < 3/7$. Applying this for $x_1 = 1$, $x_1 = 0$ next yields

$$\mu_2(1) > \frac{4}{7}; \quad \mu_2(1) + \mu_2(]\frac{1}{2}, 1[) - \mu_2(]0, \frac{1}{2}[) < \frac{3}{7} \qquad (8)$$

Next apply it to $x_1 = \frac{1}{2} - \epsilon$ and let $\epsilon > 0$ go to zero.

$$\mu_2(0) + \mu_2(1) - \mu_2(\frac{1}{2}) + \mu_2(]0, \frac{1}{2}[) - \mu_2(]\frac{1}{2}, 1[) < \frac{3}{7}$$

Summing up the last two inequalities,

$$2\mu_2(1) + \mu_2(0) - \mu_2(\frac{1}{2}) < \frac{6}{7}$$

Taking $\mu_2(1) > \frac{4}{7}$ into account, this implies $\mu_2(1/2) > (2/7)$. But (8) implies also

$$\{\mu_2(1) - \mu_2(]0, \frac{1}{2}[) < \frac{3}{7}, \ \mu_2(1) > \frac{4}{7}\} \Rightarrow \mu_2(]0, \frac{1}{2}[) > \frac{1}{7}$$

A contradiction results since $\mu_2(1) > (4/7)$, $\mu_2(1/2) > (2/7)$ and $\mu_2(]0, 1/2[) > (1/7)$.

We have just proved $\inf\limits_{\mu_2} \sup\limits_{\mu_1} \overline{u}_1 \geq (3/7)$. Equality follows by considering $\mu^*_2 = 1/7 \, \delta_{1/4} + 2/7 \, \delta_{1/2} + 4/7 \, \delta_1$. We let the reader check this last claim.

We prove next $\sup\limits_{\mu_1} \inf\limits_{\mu_2} \overline{u}_1 \leq \frac{1}{3}$ by deriving a contradition from $\inf\limits_{x_2} \overline{u}_1(\mu_1, \delta_{x_2}) > 1/3$. Taking $x_2 = 1$ this gives

$$\mu_1([0, \frac{1}{2}[) - \mu_1(]\frac{1}{2}, 1[) > \frac{1}{3} \qquad (9)$$

Taking $x_2 = 1/2 - \epsilon$ and letting $\epsilon > 0$ go to zero:

$$-\mu_1([0, \tfrac{1}{2}[) + \mu_1([\tfrac{1}{2}, 1]) > \tfrac{1}{3}$$

Summing up, $\mu_1(1/2) + \mu_1(1) > (2/3)$. Yet $\mu_1([0, 1/2[) > 1/3$ (by (9)).

To prove finally sup inf $\bar{u}_1 = 1/3$, take $\mu_1^* = 1/3\, \delta_0 +$
$\qquad\qquad\qquad\qquad\quad \mu_1 \quad \mu_2$

$1/3\, \delta_{1/2} + 1/3\, \delta_1$ and check inf $\bar{u}_1(\mu_1^*, \delta_{x_2}) = 1/3$.

We state finally the general existence result of mixed-Nash equilibria over infinite strategy sets.

Definition 2.

Let $G = \{X_i, u_i; i = 1, \ldots, n\}$, the initial N-normal-form game, be such that:

X_i is a compact subset of some
  euclidian space R^{p_i} $\qquad\qquad$ for all $i = 1, \ldots, n$
u_i is a continuous function on X_N

A mixed strategy μ_i of Player i is a Radon probability measure on X_i. It is a positive, continuous linear form on the set of continuous real functions on X_i endowed with the uniform convergence topology; its value on the 1-constant function is one. We denote by M_i the set of Player i's mixed strategies. The mixed extension of G is the game $G_m = \{M_i, \bar{u}_i; i = 1, \ldots, n\}$

$$\overline{u}_i(\mu) \;=\; \int_{X_N} u_i(x)\, d\,\mu(x) \qquad\qquad where$$

$$d\,\mu(x) \;=\; \underset{i\in N}{\otimes}\; d\,\mu_i(x_i) \qquad\qquad is\ the\ product\ of\ the$$

$$probability\ measures\ \mu_i,$$

$$i\,\epsilon\,N$$

Theorem 2 (Glicksberg [1952]).

Under the assumptions of Definition 2, the mixed game G_m has at least one Nash equilibrium outcome.

Remark 2.

For two-person, zero-sum games the assumptions of Glicksberg's Theorem can be significantly weakened. Mertens [1984] proved that if u_1 is upper semicontinuous in x_1 and lower semi-continuous in x_2 (and u_1 is bounded over $X_1 \times X_2$), then the mixed game has a value and each player has ϵ optimal strategies with a finite carrier (for all $\epsilon > 0$).

Two recent discussions of mixed-Nash equilibria in infinite games have been made by Parthasarathy-Raghavan [1971] and Tijs [1981].

5. EXERCISES

a) Finite two-person, zero-sum games

1) Let $B = \begin{pmatrix} 21 \\ 45 \end{pmatrix}$. Player 1 chooses either a row of B or a column of B. Player 2 chooses an element of (position in) B. If the element picked by Player 2 is in the row or column

chosen by Player 1, then Player 2 pays Player 1 that amount. Otherwise Player 2 pays Player 1 zero. Solve this game.

2) One-card, one-round bluffing

With equal probability Player 1 is dealt card H or card L. Player 2 is not dealt a card, and never gets to look at Player 1's card until the end of the game. After looking at his card, Player 1 either plays or folds. If he folds, he loses $1. If he plays, Player 2 now must consider if he should fold. If he decides to fold, he loses $1. If he sees, the card is shown. If it is H, Player 1 wins $4 from Player 2; if it is L, Player 2 wins $a from Player 2.

Draw a game tree and the normal form of the game. Show that Player 1 has only two undominated strategies. Find the mixed value and optimal strategies of this game as a function of a (restrict yourself to the case a > 1).

3) One-card, two-round bluffing

After looking at his card, Player 1 either plays or folds. If Player 1 folds, he loses $2. If Player 1 plays, Player 2 now must raise or fold. If Player 2 folds, he loses $2. If Player 2 raises, then Player 1 must again either play or fold. If Player folds he loses $4. If Player 1 plays, Player 2 must now either fold or see. If Player 2 folds, he loses $4. If Player 2 sees, Player 1 must show the card. If it is H, Player 1 wins $6; if it is L, Player 1 loses $6. Formally, Player 1 has nine pure strategies and Player 2 has three (Why?).

Reduce Player 1's strategy set to three strategies by domination arguments. Then solve the mixed extension of that game.

4) Solve the recursive two-person, zero-sum game:

$$ G = \begin{bmatrix} 0 & 1 \\ 1 & -G/2 \end{bmatrix} $$

Make up a brief scenario for G.

5) <u>Another inspector-cheater game</u>

There are n rounds in this game involving a cheater and an inspector. The cheater may cheat in every round. He gains \$1 for every round in which he cheats without being inspected. If he is inspected, he is fined \$d and is "put out of business." That is, he cannot cheat again. The inspector can inspect only once, and this inspection is "noisy." That is, the cheater knows when it happens. Express this as a recursive game. That is, give G_1 (one round left) explicitly and G_n in terms of G_{n-1}. Check that at round n, the optimal strategy has to be mixed and deduce an induction formula for v_n, the mixed value of G_n, as a function of v_{n-1}. Solve and describe both players' optimal strategies.

6) <u>Head-tail-head</u>

Each player chooses a triple $\alpha_1 \alpha_2 \alpha_3$ where each α_i is either a head or tail. Thus each has eight different strategies. Then a coin is tossed until one player's figure appears. This

181

player is paid \$1 by the other (of course, if both players choose the same figure, the game is a draw, and no coin has to be tossed).

For instance, take x_1 = TTH, x_2 = THH. If the drawing gives HTHTTH ..., it is a win for Player 1. In fact, given these strategies, the expected payoff to Player 1 is worth 1/3, for after T is drawn and the next one is T, Player 1 wins for sure. However, if the next draw is H, there is a 1/2 probability that Player 2 will win (if the next is H) and a 1/2 probability of landing back at the starting point (if the next draw is T). Show that similar computations for each pair of strategies lead to the accompanying payoff matrix.

	HHH	HHT	HTT	HTH	THT	THH	TTH	TTT
HHH			-1/5	-1/5	-1/6	-3/4	-2/5	
HHT			1/3	1/3	1/4	-1/2		2/5
HTT	1/5	-1/3					1/2	3/4
HTH	1/5	-1/3					-1/4	1/6
THT	1/6	-1/4					-1/3	1/5
THH	3/4	1/2					-1/3	1/5
TTH	2/5		-1/2	1/4	1/3	1/3		
TTT		-2/5	-3/4	-1/6	-1/5	-1/5		

Note that blank entries are zero. Also, by symmetry arguments only 1 pair of strategies are nontrivial. Eliminate dominated strategies, then find optimal mixed strategies.

7) Player 2 chooses one of the three numbers 1, 2, 5 , say, x_2. One of the two numbers not chosen is selected at random and shown to Player 1. Player 1 now guesses which number Player 2 chose, winning $\$x_2$ from him if he is correct. If Player 1 guesses incorrectly the payoffs are zero to both.

Find the mixed value of this game and optimal strategies for both players.

Hint: It is cumbersome to draw the 8 x 3 normal form of this game. Use instead the following description of Player 1's mixed strategies, called "behavioral strategies." If he is shown number 1, he guesses 2 with probability q_1 and 5 with probability $1 - q_1$, etc. Player 1's strategy is now a triple (q_1, q_2, q_5) where each q_k is in $[0, 1]$. Write the corresponding normal form and solve.

8) Give an example of a 2 x 2 zero-sum game where Player 1 has a completely mixed strategy μ_1^* such that $u_1(\mu_1^*, \mu_2)$ is independent of $\mu_2 \in M_2$, yet μ_1^* is not optimal in the mixed game. Find a two-person, zero-sum game with no value (in pure strategies) where the same situation arises.

9) Let $G = (X_1, X_2, u_1)$ be a finite two-person, zero-sum game where no player has an optimal completely mixed strategy.

For all $(x_1, x_2) \in X$ denote $v_1(x_1, x_2)$ the mixed value of the game $(X_1 \setminus \{x_1\}, X_2 \setminus \{x_2\}, u_1)$. Show that the game (X_1, X_2, v_1) has a saddle point in pure strategies and its value is the mixed value of G.

10) If the finite two-person, zero-sum game $G = (X_1, X_2, u_1)$ has a value (in pure strategies), prove that any mixed strategy in G_m, the carrier of which is made solely of optimal strategies of G, is optimal in G_m. Given an example where G_m has more optimal strategies.

11) Let (X_1, X_2, u) be a finite two-person, zero-sum game with no value in pure strategies.

$$\sup_{x_1} \inf_{x_2} u = \alpha_1 < \alpha_2 = \inf_{x_2} \sup_{x_1} u$$

a) Suppose that for each x_1 (respectively for each x_2) the mapping $x_2 \to u(x_1, x_2)$ is one to one on X_2 (respectively $x_1 \to u(x_1, x_2)$ is one to one on X_1). Prove that the mixed value $v_m(u)$ strictly improves the secure utility level of both players

$$\alpha_1 < v_m(u) < \alpha_2$$

b) Give an example where the one to one assumption is false and

$$\alpha_1 = v_m(u) < \alpha_2$$

(Hint: a 3 x 2 game will do).

b) Finite n-person games

12) Consider the 2 x 2 game

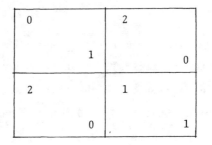

Compute its (unique) mixed Nash equilibrium (μ^e_1, μ^e_2) and its
(unique) prudent mixed strategies (μ^p_1, μ^p_2). Assuming that
each player is hesitating between μ^e_i and μ^p_i, draw the
corresponding 2 x 2 game and analyze it. Use this to give
some strategical advice to each player.

13) Give an example of a 2 x 2 game where i) no player has
two equivalent pure strategies, ii) there is an infinite
number of Nash equilibria.

14) To $G = (X_1, X_2, u_1, u_2)$ associate two $|X_1| \times |X_2|$ matrices
$U_i (i = 1, 2)$:

$$U_i = [u_i(x_1, x_2)] \quad x_1 \in X_1$$
$$x_2 \in X_2$$

Suppose both U_1 and U_2 can be inverted, so that $|X_1| = |X_2|$ and $\det(U_i) \neq 0$ and denote by $\tilde{U}_i(x_1, x_2)$ the cofactor of entry (x_1, x_2) in matrix U_i. Prove that if G has a completely mixed Nash equilibrium, the corresponding payoff to Player i is

$$\frac{\det (U_i)}{\displaystyle\sum_{\substack{x_1 \in X_1 \\ x_2 \in X_2}} \tilde{U}_i(x_1, x_2)}$$

15) Where Player 1 gains by losing utility (Moulin [1976])

In the first version of the crossing game (Example 6) show that Player 1's payoff from a mixed Nash equilibrium is increased (and Player 2's payoff is unchanged) if Player 1 is penalized (by a small enough amount) for using strategy go. In other words, prove the following.

	1		1 - ε	
		1		2
G =	2		0	
		1 - ε		0

	1		1 - ε	
		1		2
G' =	2 - α		-α	
		1 - ε		0

For $\varepsilon > 0$, $\alpha > 0$ both small enough, the mixed Nash-equilibrium payoffs in G, G' are such that

$$v_1 < v'_1 \; ; \quad v_2 = v'_2$$

16) In a certain symmetric three-person game, each player has pure strategies A and B. If Players 1, 2, and 3 play the mixed strategies $(p, 1 - p)$, $(q, 1 - q)$ and $(r, 1 - r)$ respectively, the payoff to Player 1 is

$$u_1(p, q, r) = pqr + 3pq + pr + qr - 2q - p$$

Since the game is symmetric, there is an equilibrium in which all three players play the same mixed strategy. Find this strategy.

17) Exchangeability of mixed NE outcomes in two-person games
(Parthasarathy-Raghavan [1971])

In a two-person mixed game $G_m = (M_1, M_2, \bar{u}_1, \bar{u}_2)$ the NE outcomes usually are not exchangeable.

$$\mu, \mu' \in NE(G_m) \not\Rightarrow (\mu_1, \mu'_2), (\mu'_1, \mu_2) \in NE(G_m)$$

This should be clear from the crossing game (Example 6).
a) Prove that the NE outcomes of G_m are exchangeable iff $NE(G_m)$ is a convex subset of $R^{X_1} \times R^{X_2}$ (where X_i are, of course, finite). b) If $NE(G_m)$ is indeed a convex (hence rectangular) subset of $M_1 \times M_2$, prove the existence of at least one mixed NE outcome μ^* that Pareto dominates or has the same payoff vector as all other mixed NE outcomes.

$$\forall\, \mu \in NE(G_m) \quad u_i(\mu) \le u_i(\mu^*) \qquad\qquad i = 1, 2$$

c) Infinite games

18) Players 1 and 2 each pick a positive integer. The player
with the lower integer wins \$1 from the other unless he is
exactly one lower, in which case he loses \$2. If they both
pick the same number, the payoff is zero to both. Find the
unique optimal mixed strategy (the value is zero by symmetry).
Hint: The carrier of an optimal strategy is finite.

19) Hiding a number

Two players pick a positive integer. If these two numbers
differ, the payoff is zero. If they coincide, $x_1 = x_2 = p$,
Player 1' receives a_p from Player 2 where $a_1 \leq a_2 \leq \ldots \leq a_p \leq \ldots$
is a nondecreasing sequence of positive numbers. If they do not,
the payoff is zero to both. Thus, Player 2 faces a dilemma.
The larger his number, the less probable it is that Player 1 can
guess it, but the more risky a potential discovery. a) Write
the normal form of the pure and mixed games. Observe that the
payoff might be $+\infty$. b) Suppose first

$$\sum_{p=1}^{+\infty} \frac{1}{a_p} = \frac{1}{\alpha} < +\infty$$

Prove that there is a unique, symmetric, completely mixed
saddle point and the value is α. c) Suppose next

$$\sum_{p=1}^{+\infty} \frac{1}{a_p} = +\infty$$

Prove that the value is zero, that every strategy of Player 1 is optimal, and yet Player 2 has no optimal strategy. Describe an ε optimal strategy for Player 2.

20) Catch me

a) Player 2 picks a number x_2 in $[0, 1]$. Player 1 tries to catch him by picking an x_1 in $[0, 1]$ within a distance 1/4:

$$X_1 = X_2 = [0, 1] \quad u_1(x_1, x_2) = 1 \quad \text{if } |x_1 - x_2| \leq \frac{1}{4}$$

$$= 0 \quad \text{if } |x_1 - x_2| > \frac{1}{4}$$

In the pure-strategy game, we have $\alpha_1 = 0 < 1 = \alpha_2$. Prove that the mixed extension of the game has value 1/2 and compute a pair of optimal strategies.

Hint: Seek optimal mixed strategies where each player uses only two pure strategies.

b) Consider now the variant of the game of catch me where

$$u_1(x_1, x_2) = 1 \qquad \text{if } |x_1 - x_2| < \frac{1}{4}$$

$$= 0 \qquad \text{if } |x_1 - x_2| \geq \frac{1}{4}$$

Although the change in the payoff function is small, the value decreases to 1/3. Explain why by giving a pair of optimal mixed strategies.

21) First-price, auction dollar game

In a variant of the auction dollar game (Example 2, Chapter 2)

the highest bidder gets the prize and every player pays his
own bid:

$$u_i(x) = 1 - x_i \qquad \text{if } x_i > x(i)$$

$$= -x_i \qquad \text{if } x_i < x(i)$$

where $x(i) = \sup\limits_{j \neq i} x_j$.

Prove that every strategy $x_i > 1$ is dominated. Then compute
the symmetrical mixed-Nash equilibrium with a positive density
on $[0, 1]$.

Hint: See Example 7.

22) A location game

Two shopowners must decide the location of their respective
shops along an interval $[0, 1]$. Player 1 sells cheap sports
equipment, whereas Player 2 deals in elegant sports gear.
As side-by-side comparison is, on average, favorable to the
merchant of cheap equipment, the players face an inelastic
demand; Player 1 wants to locate as close as possible to
Player 2, whereas Player 2 tries to move as far as possible
from Player 1. We assume that the profit functions take the
following form.

$$X_1 = X_2 = [0, 1]$$

$$u_1(x_1, x_2) = 1 - |x_1 - x_2|$$

$$u_2(x_1, x_2) = |x_1 - x_2| \qquad \text{if } |x_1 - x_2| \leq \tfrac{2}{3}$$

$$= \tfrac{2}{3} \qquad \text{if } |x_1 - x_2| \geq \tfrac{2}{3}$$

Notice the negative externalities imposed by Player 1 on
Player 2 vanish when their distance is at least 2/3.

The pure-strategy game has no Nash equilibrium. Show that
Glicksberg's theorem implies the existence of a mixed Nash
equilibrium. Check that the following pair is a Nash equilibrium.

$$\mu^*_1 = \frac{1}{3}\delta_0 + \frac{1}{6}\delta_{\frac{1}{3}} + \frac{1}{6}\delta_{\frac{2}{3}} + \frac{1}{3}\delta_1$$

$$\mu^*_2 = \frac{1}{2}\delta_0 + \frac{1}{2}\delta_1$$

Neither of these two mixed strategies are completely mixed.
However (μ^*_1, μ^*_2) share the typical property of completely
mixed equilibria, namely,

$$\bar{u}_1(\mu_1, \mu^*_2) = \bar{u}_1(\mu^*_1, \mu^*_2) \qquad \text{all } \mu_1 \in M_1$$

$$\bar{u}_2(\mu^*_1, \mu_2) = \bar{u}_2(\mu^*_1, \mu^*_2) \qquad \text{all } \mu_2 \in M_2$$

PART III.   COOPERATIVE SCENARIOS

CHAPTER 8.   CORRELATED EQUILIBRIUM

1.   CORRELATED EQUILIBRIUM A LA AUMANN

In parts I and II we ruled out overt communication among players, as a result of physical, legal barriers and/or mutual distrust. Mutual observations of tactical moves were the only, indirect, way of conveying some strategical information (see the Cournot tatonnement, Chapter 6). Now, we seek to design behavioral patterns of a cooperative society, where open communication takes place. Structural inefficiency of the noncooperative equilibrium outcome (an unambiguous feature in most nonzero-sum games) is now interpreted as an incentive to cooperate. We attempt to describe its likely manifestations.

After cooperative communication, which may involve mutual disclosure of the utility functions, bargaining, and various psychological maneuvers, the players reach an agreement. Agreements can be binding (when several players can sign a contract to use specified strategies, and this contract can be enforced by some authority to whom obedience is due by all players) or nonbinding, when no such authority exists so that an agreement, like most international treaties, is respected as long as betrayal does not pay. In this book we study nonbinding agreements only, for we feel that the subject of binding agreements raises the mainly normative issue of equity (which outcome is fair agreement?) to which strategical analysis contributes very little. The Shapley value of a game in characteristic form is the typical example of an equity concept with no strategical content. We shall not discuss it. The interested reader should consult Aumann [1976] or Owen [1982].

A nonbinding agreement specifies an outcome, that is, a particular strategy for each player. We assume each player keeps full sovereignty over his or her strategic variable, so that no player nor any coalition of players can be forced to use the agreement strategy. Thus, the only way to prevent betrayal by an individual agent or a coalition of agents is to make this deviation unprofitable to the deviants. This is the self-enforcing property.

The fundamental example of a self-enforcing agreement is that of a Nash equilibrium, the stability of which is enhanced by mutual secrecy concerning the final strategic decision. Mixed strategies are useful because they raise new Nash equilibria. From a cooperative point of view, a mixed Nash equilibrium is a nonbinding agreement self-enforced by the mutual privacy of the lotteries that each player uses to pick at random his final strategy.

Here we explore two cooperative variants of the Nash-equilibrium concept. We assume that our players can randomly correlate their strategies, and send back a confidential signal to each individual. This in turn generates new self-enforcing agreements.

Example 1. Another crossing game.

Consider the following variant of the crossing game (Example 6, Chapter 7). In addition to its two Nash equilibria in pure strategies, the game has a symmetrical equilibrium in completely mixed strategies. Both stop with probability 5/8, and the expected payoff is .81 for both.

	Stop	Go
Stop	1 1	.5 1.3
Go	1.3 .5	0 0

A correlated equilibrium results if the players agree to
build red lights, namely, a random signal showing (green, red)
or (red, green) with equal probability. The agreement is
made that whoever sees green goes, whereas whoever sees red
stops. This agreement is self-enforcing since, for each
drawing of the lottery, the corresponding outcome agreed upon
is a Nash equilibrium. The resulting expected payoff is .9,
thereby improving upon the mixed Nash equilibrium.

But they can do better. Suppose the random signal is
built up so as to show (green, red) with a probability of .4,
(red, green) with a probability of .4 and (red, red) with a
probability of .2. Suppose, moreover, that each player
observes only his or her own coordinate. Player 1 is told the
left coordinate, and Player 2 is told the right one. Thus,
when Player 1 sees green, he knows for sure that Player 2 sees
red. If he sees red, he knows that Player 2 sees green with
a probability of 2/3 or red with a probability of 1/3. This
skillful uncertainty is enough to give him every incentive to
stop whenever he sees red, assuming that Player 2 will go if
he sees green and will stop if he sees red. Player 1's
expected payoff while stopping is then 1/3(1) + 2/3(.5) = .66
which exceeds his expected payoff while going 1/3(1.3) + 2/3(0) =
.43. Of course, if Player 1 sees green, he has, as before,
every incentive to go, assuming that Player 2 obeys the red
signal. Thus the agreement to abide by the color that is seen
if self-enforcing. The expected payoff is now .92 to both.

Definition 1 (Aumann [1974]).

Given a game $G = (X_i, u_i, i = 1, \ldots, n)$ with finite strategy sets, we denote by $L = (L(x))_{x \in X_N}$ a _correlated lottery_, that is, a probability distribution over X_N. For all $i \in N$, and all $x_i \in X_i$ we denote by L_{x_i} the conditional probability of L over X_{-i}:

$$\text{all } x_{-i}, \quad L_{x_i}(x_{-i}) = \frac{1}{\sum_{y_{-i} \in X_{-i}} L(x_i, y_{-i})} \cdot L(x_i, x_{-i})$$

if the denominator is nonzero.

$$= 0 \text{ if } L(x_i, y_{-i}) = 0$$

for all $y_{-i} \in X_{-i}$.

We say that L is a _correlated equilibrium_ of G if the following inequalities hold.

$$\forall i \in N \ \forall x_i, y_i \in X_i \quad \sum_{x_{-i} \in X_{-i}} u_i(x_i, x_{-i}) L_{x_i}(x_{-i}) \geq$$

$$\sum_{x_{-i} \in X_{-i}} u_i(y_i, x_{-i}) L_{x_i}(x_{-i})$$

(1)

We denote by $CE(G)$ the set of correlated equilibria.

The cooperative scenario justifying Definition 1 is the following. The players cooperatively construct a random signal that potentially selects outcome $x \in X_N$ with probability

$L(x)$. If outcome x is drawn, <u>Player i is informed of the</u>
<u>component x_i only</u>; then each player selects independently and
secretly his or her actual strategy. The signal x_i to
Player i is understood as a nonbinding suggestion to play
x_i. Condition (1) says that the agreement to obey this
suggestion is self-enforcing, given the limited information
available to each player. Suppose Agent i is told by the
signal to use strategy x_i. He infers from the overall
distribution L that the $N\backslash\{i\}$-tuple x_{-i} has been selected with
probability $L_{x_i}(x_{-i})$. Therefore,

$$[u_i(y_i, \cdot), L_{x_i}] = \sum_{x_{-i} \in X_{-i}} u_i(y_i, x_{-i}) L_{x_i}(x_{-i})$$

is the expected utility of playing strategy $y_i \in X_i$ given that
all other players obey the signal. Thus, Condition (1) means
that if the strategy suggested by the signal is used, it always
is optimal to Player i given the information available to him
and the assumption that the other players obey the signal.

Suppose that x_i is such that $L(x_i, x_{-i}) = 0$ for all
$x_{-i} \in X_{-i}$; i.e., the probability that x_i is ever suggested by
the signal is zero. For such an x_i, Condition (1) is trivially
satisfied, hence system (1) is equivalent to

$$\forall \, i \in N \, \forall \, x_i, \, y_i \in X_i \quad \sum_{x_{-i} \in X_{-i}} u_i(x_i, x_{-i}) L(x_i, x_{-i}) \geq$$

$$\sum_{x_{-i} \in X_{-i}} u_i(y_i, x_{-i}) L(x_i, x_{-i})$$

(4)

From this it follows that a lottery L is a correlated equilibrium iff it satisfies a system of <u>linear</u> inequalities. Actually this system always has a solution.

<u>Lemma 1.</u>

Notations and assumptions are the same as those used in Definition 1. The set CE(G) of correlated equilibria of G is a nonempty, convex compact subset of the unit simplex of R^{X_N}. If $\mu = (\mu_i)_{i \in N}$ is an outcome of the mixed game G_m, then the associated lottery $L = \underset{i \in N}{\otimes} \mu_i$

$$L(x) = \prod_{i \in N} \mu_i(x_i) \qquad (2)$$

is a correlated equilibrium of G iff μ is a Nash equilibrium of G_m.

<u>Proof.</u>

Let $\mu = (\mu_i)_{i \in N}$ be an outcome of the mixed game G_m and L be the corresponding product lottery given by (2). System (1) becomes

$$\mu_i(x_i) \cdot \bar{u}_i(\delta_{x_i}, \mu_{-i}) \geq \mu_i(x_i) \cdot \bar{u}_i(\delta_{y_i}, \mu_{-i}) \qquad (3)$$

for all i, all x_i, y_i. The inequality is trivial when $\mu_i(x_i) = 0$. Hence (3) is equivalent to

$$\bar{u}_i(\delta_{x_i}, \mu_{-i}) \geq \bar{u}_i(\delta_{y_i}, \mu_{-i})$$

for all $i \in N$, all $x_i \in [\mu_i]$, and all $y_i \in X_i$. \qquad (4)

Suppose now that μ is a mixed NE outcome of G. Then, by Lemma 4, Chapter 7, system (4) holds. Conversely, (4) implies that $\bar{u}_i(\delta_{x_i}, \mu_{-i})$ does not depend on $x_i \in [\mu_i]$ and is therefore equal to $\bar{u}_i(\mu)$. This concludes the proof of the second statement.

Next from Nash's theorem (Theorem 2, Chapter 5) the set $NE(G_m)$ is nonempty, which implies the nonemptyness of CE(G). Convexity and compactness of the latter set follow from our remark that CE(G) is defined by a family of linear closed inequalities. QED.

The argument justifying that a correlated equilibrium outcome is a self-enforcing agreement is identical to that supporting the cooperative view of the Nash-equilibrium outcomes. Actually by Lemma 1 a NE outcome, whether of the initial game or of its mixed version, is identified with a correlated equilibrium L where the probability distribution L is that of a N-tuple of independent random individual strategies. In that case, really no correlation of the strategies of the various players takes place.

One first, easy way to achieve some correlation of individual strategies is to take a convex combination of NE outcomes. Typically a lottery L is chosen such as

$$L = \sum_{\alpha=1}^{p} \lambda_\alpha \delta_{x_\alpha}, \quad \sum_{\alpha=1}^{p} \lambda_\alpha = 1$$

$\lambda_\alpha \geq 0$ for all α, where δ_x is the lottery with weight 1 on outcome x, and x is a (pure) NE outcome of G. Then L is a

correlated equilibrium of G (notice that convex combinations

of mixed NE outcomes also are in CE(G)). In many games,

however, the set of correlated equilibrium outcomes is bigger

than the mere convex hull of Nash equilibria (see Example 1).

Our next example stresses this important point.

Example 2. Musical chairs.

In this nonstandard version of musical chairs, we have

two players and three chairs marked 1, 2, 3. Each player's

strategy is to pick a chair. Everybody loses if both players

pick the same chair. If, on the contrary, the two choices

differ, this player — say i — whose chair immediately follows

that of j wins twice more than j (with the convention that

1 follows 3). Hence the bimatrix game

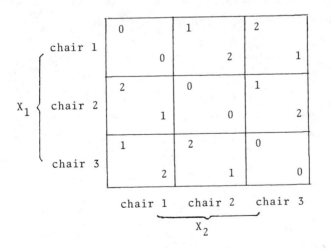

(5)

The initial game has no Nash-equilibrium outcome.
There is a unique, completely mixed, Nash equilibrium, namely,

$$\mu^*_1 = \mu^*_2 = \frac{1}{3}\delta_1 + \frac{1}{3}\delta_2 + \frac{1}{3}\delta_3$$

This symmetrical outcome yields payoff 1 to both players
and is Pareto dominated. This is so because in (μ^*_1, μ^*_2) a
bad deterministic outcome (i, i) is drawn with probability
1/3. Consider now the following lottery L over $X_1 \times X_2$,

$$L(x_1, x_2) = \frac{1}{6} \qquad \text{if } x_1 \neq x_2$$
$$= 0 \qquad \text{if } x_1 = x_2$$

We claim that L is a correlated equilibrium of game (5).
Suppose, for instance, that the deterministic outcome (2, 3)
has been selected and Player 1 is therefore told to use
strategy 2. Given L, Player 1 infers from this message that
Player 2 is told to use either one of the strategies 1 or 3,
each one of these with probability 1/2. In other words,
assuming that Player 2 obeys, the signalled strategy amounts
to player 1's assuming that Player 2 uses the mixed strategy
$\mu_2 = (1/2) \delta_1 + (1/2) \delta_3$. To the latter, his best reply
actually is to use strategy 2 since

$$\bar{u}_1(\delta_2, \mu_2) = \frac{3}{2} > \bar{u}_1(\delta_1, \mu_2) = 1 > \bar{u}_1(\delta_3, \mu_2) = \frac{1}{2}$$

Similarly, Player 2 infers from the current signal (play
strategy 3) that the signal to Player 1 is strategy 1 with

201

probability 1/2 or strategy 2 with probability 1/2. Next

Player 2's best reply to Player 1's mixed strategy

$\mu_1 = (1/2) \delta_1 + (1/2) \delta_2$ happens to be strategy 3.

The payoff vector of L, namely (3/2, 3/2), is Pareto

optimal (even within the set of correlated lotteries).

2. WEAK CORRELATED EQUILIBRIUM

In our next example, correlated equilibria do not allow

any improvement of the unique Nash equilibrium. Nonetheless,

a more-stringent form of self-enforcing agreement, relying

on correlation again, is the accurate cooperative device.

Example 3. Competition by differentiation.

Two duopolists compete by choosing the quality of the

good they supply on the same market. The game is symmetrical

and each firm can choose among three strategies: low, medium,

or high quality. If both supply a low-quality product or

both supply a high-quality one, they both receive a zero profit.

If one player offers medium quality whereas the other is at

high or low, the medium supplier of the medium-quality product

gets a profit of 2, whereas the other gets nothing. If both

goods are medium-quality, each player received 1. To achieve

the maximal joint profit the two firms must offer one high-

quality good and one low-quality product, in which case the

producer of high-quality products receives 3, whereas the

manufacturer of low-quality products receives 1.

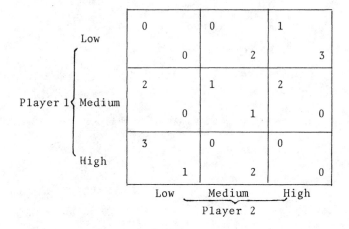

For noncooperative players, the game is dominance solvable. Strategies to produce low quality are dropped first, next strategies high. This implies that (M, M) is the unique NE outcome of the initial game as well as its mixed extension. Also, the unique correlated equilibrium outcome is the lottery with mass 1 on (M, M). (To check this claim, remark that a CE lottery gives weight zero to any strictly dominated strategy. Then apply this remark twice.)

However the Pareto optimal payoff vector (2, 2) can be achieved as follows. Construct a lottery with probability distribution

$$L = \frac{1}{2} \delta_{(H, L)} + \frac{1}{2} \delta_{(L, H)}$$

Each player, secretly and independently, sends a <u>binding</u> signal s_i to a neutral referee. Four signals are available to each player: the three pure strategies and signal OB (obedience to the lottery). After a pair (s_1, s_2) of messages are sent, the referee draws at random an outcome (x_1, x_2) according to lottery L. The final outcome is given by the following rule (enforced by the referee).

(x_1, x_2) if $s_1 = s_2 = OB$

(x_1, s_2) if $s_1 = OB$, $s_2 = L, M, H$

(s_1, x_2) if $s_1 = L, M, H$, $s_2 = OB$

(s_1, s_2) if $s_i = L, M, H$ $i = 1, 2$

In words, signal OB is a commitment to play that strategy selected by the lottery, taken by a player before the lottery is drawn. A signal like M is just the usual pure strategy M. Here, the pair of obedient signals is a self-enforcing agreement, namely,

$$\bar{u}_1(OB, OB) = \frac{1}{2}u_1(H, L) + \frac{1}{2}u_1(L, H) = 2 > \bar{u}_1(y_1, OB) = \frac{1}{2}u_1(y_1, L)$$
$$+ \frac{1}{2}u_1(y_1, H)$$

for all y_1 in $\{L, M, H\}$.

$$\bar{u}_2(OB, OB) = \frac{1}{2}u_2(L, H) + \frac{1}{2}u_2(H, L) = 2 > \bar{u}_2(OB, y_2) = \frac{1}{2}u_2(H, y_2)$$
$$+ \frac{1}{2}u_2(L, y_2)$$

for all y_2 in $\{L, N, H\}$.

Contrary to correlated equilibria, the decision to obey the lottery cannot be revised after a particular outcome has been drawn. It must be made once and for all before the random drawing.

Definition 2 (Moulin-Vial [1978]).

Notations and assumptions are the same as those used in Definition 1. For all $i \in N$, we denote by L_{-i} the marginal distribution of L over x_{-i}, namely,

$$L_{-i}(x_{-i}) = \sum_{x_i \in X_i} L(x_i, x_{-i})$$

for all x_{-i}.

We say that lottery L is a weak correlated equilibrium of G if the following inequalities hold.

$$\forall i \in N \; \forall y_i \in X_i \quad \sum_{x \in X_N} u_i(x) L(x) \geq \sum_{x_{-i} \in X_{-i}} u_i(y_i, x_{-i}) L_{-i}(x_{-i})$$

We denote by $WCE(G)$ the set of weak correlated equilibria of G.

Lemma 2.

The set $WCE(G)$ is a nonempty convex compact subset of the unit simplex of R^{X_N}. It contains the set $CE(G)$ of correlated equilibria of G as a subset.

If μ is an outcome of the mixed game G_m, then the associated lottery $L = \bigotimes_{i \in N} \mu_i$ is a weak correlated equilibrium of G iff μ is a Nash equilibrium of G_m.

Proof.

By Definition 2, lottery L belongs to WCE(G) iff

$$\forall \ i \ \epsilon \ N \ \forall \ y_i \ \epsilon \ X_i \sum_{x \epsilon X_N} u_i(x) L(x) \geq \sum_{x \epsilon X_N} u_i(y_i, x_{-i}) L(x)$$

These inequalities are obtained from system (1) by keeping i and y_i fixed while summing up over $x_i \ \epsilon \ X_i$. Hence the inclusion

$$CE(G) \subset WCE(G) \qquad \underline{QED}.$$

Definition 2 represents the last step in the conceptual development of self-enforcing agreements using randomization and mutual secrecy. Starting with the Nash equilibrium in pure strategies, we allow first for independent randomizations of individual strategies — mixed strategies — and observe that if the players can draw these private lotteries in mutual secrecy, self-enforcing agreements always exist (Theorem 1, Chapter 7). Next, if the players can build correlated lotteries sending decentralized signals to individuals, the set of equilibria becomes convex. From Lemma 1 and 2 we deduce the inclusions

$$NE(G) \subset NE(G_m) \subset CE(G) \subset WCE(G)$$

From left to right more and more informational constraints must be met to make an equilibrium outcome self-enforcing. For a NE or a mixed NE outcome, only the privacy of individual strategies has to be preserved. For a CE outcome, we must,

in addition, prevent individual players from observing any-
thing but their own component of the actual drawing of the
correlated lottery. For a WCE outcome, a neutral referee
must draw an outcome of which <u>nothing</u> is revealed to any
player, next ask each player independently and secretly
whether or not he is willing to play this "blind" outcome,
inform only those players who voluntarily choose to obey the
lottery of the actual drawing of the lottery, and force the
declared obedient players to play the required strategy
faithfully. In short, the referee must have the power to
block any strategic use of the actual drawing of the lottery.

The common feature of these informational scenarios is
that after an agreement has been reached on the choice of a
particular correlated strategy, no more direct communication
can take place among players. In the case of a repeated
drawing of the agreed-upon lottery, cooperation becomes implicit
and is hardly recognizable. This form of tactical collusion
is attested by the literature on oligopolistic behavior. One
example of such behavior was apparent in the "electrical
equipment conspiracy of the 1950s" (Sherer [1970]) which involved
29 U.S. companies selling different sorts of products. For
the power-switching equipment, sealed-bid competitions were
sponsored by the U.S. government. Through some preliminary
secret negotiation, each bidder was assigned a share of all
the business. The sellers then coordinated their bidding
so that each one of them was low bidder in just enough

transactions to gain its predetermined share of the market. This was achieved by dividing the territory into four quadrants assigning different sellers to each quadrant and letting the sellers in a quadrant rotate their bids. A "phases of the moon" system was used to decide which company could make the low-bid. Thus, the result was an ostensibly random process conveying the impression of independent behavior (See Gerard-Varet and Moulin [1978]).

3. EXERCISES

1) In the crossing game of Example 1 find the best symmetrical correlated equilibrium.

2) In a 2 x ... x 2 game (each player has exactly two strategies) prove that Definition 1 and 2 coincide.

3) In the musical-chair game (Example 2) prove that the sets $NE(G)$, $NE(G_m)$, $CE(G)$ and $WCE(G)$ are all different. Prove that in a Pareto-optimal CE payoff, a player receives no more than 5/3 and no less than 4/3 (and both bounds are reached). However, in a Pareto-optimal WCE a player's payoff can be anywhere between 1 and 2.

CHAPTER 9.   COALITIONAL STABILITY: THE CORE

1.   IMPUTATIONS AND DETERRING SCENARIOS

Deterrence is a powerful means to induce cooperation. To enforce the stability of an agreed-upon outcome, our players threaten each other, i.e., announce a specific reaction to potential deviations. If the deviant player happens to be worse off after the announced threat has been carried out, he is deterred from deviating and the nonbinding agreement proves to be stable. So, deterrence is the "skillful exploitation of potential forces." A successful threat is one that is not carried out (Schelling [1971]).

Enforcing cooperation by means of mutually deterring threats is an old topic in the oligopoly literature. For instance, in a Cournot dupoly model (such as Example 3, Chapter 5) the maximal joint-profit outcome is not a Nash equilibrium since by unilaterally increasing its own supply a firm improves upon its instantaneous profit. However, this move results in a similar improvement for the other firm that eventually leaves both players worse off than they were with the initial outcome. As Sherer [1970] points out, "Each would be reluctant to take measures, which when countered, would leave all members of the industry worse off."

Whereas the self-enforcing agreements of Chapter 8 require strict privacy of the final strategic decision, on the contrary, a threat is effective only if any deviation is publicly known. Henceforth, stability by deterrence demands that individual strategic choices are obvious to all. Military and economic examples of the limitation of strategic privacy for the sake of cooperative stability abound (see below).

Example 1. Cooperative solution of the prisoners' dilemma.

The noncooperative outcome of the prisoners' dilemma (Example 2, Chapter 3) is war: (A_1, A_2). To enforce the cooperative peaceful outcome (P_1, P_2), both players announce the following tit-for-tat behavior.

if you play peacefully, I shall play peacefully too

$$(1)$$

if you play aggressively, I shall play aggressively too

Given that my opponent threatens me in this way, I determine that I had better be peaceful (leading to peace (P_1, P_2)) rather than aggressive (leading to war (A1. A$_2$)). In general, stability by threats is the combination of an agreed upon-outcome and a threat that deters each individual player from betraying the agreement.

Definition 1.

Let $G = \{X_i, u_i; i = 1, \ldots, n\}$ be a game. A *deterring scenario* is a $(n + 1)$-tuple $(x, \xi_{-i}; i = 1, \ldots, n)$ where $x \in X_N$ is an outcome of G and, for all $i \in N$, ξ_{-i} is a threat against Player i, that is, a mapping from X_i into X_{-i}, such that

$$\xi_{-i}(x_i) = x_{-i}$$

(2)

$$\forall \; y_i \in X_i \setminus \{x_i\}: \quad u_i(y_i, \xi_{-i}(y_i)) \leq u_i(x)$$

We interpret a deterring scenario in the framework of a game played over an infinite time period. At one particular date each player plays a particular strategy; he can change at will his strategy at any time. The strategies of all players are constantly public. This is the crucial informational feature that prevents any player from deviating secretly. A player faithful to the agreement first plays his agreed-upon strategy x_i and observes the current strategies y_{-i} used by other players. As long as $y_{-i} = x_{-i}$, Player i keeps strategy x_i; as soon as some individual player, say j, switches to strategy $y_j \neq x_j$, then i plays the i-th component of $\xi_{-j}(x_j)$

forever. Condition (2) is the self-enforcing property, given
that the payoff over an infinite period of time outweighs
that over any finite period. This interpretation is fairly
loose. We postpone until Chapter 10, Section 2 the rigorous
presentation of the repeated-game concept that justifies the
scenario above.

Surely, the informational constraint that all strategic
moves must occur publicly might be difficult to achieve. For
instance, modern armament technology makes surprise attack
more and more frightening. Thus, a demilitarized noman's land
on which all aggressive moves are visible, or a nuclear-site
inspection agreement, are both informational devices pertaining
to deterring threats. As another example, on "open price
agency" prevents competing firms from offering secret discounts
to their customers and therefore reduces the overall competition
(Sherer [1970]). An example in the opposite direction is the
Anglo-Japanese agreement to limit the production of war
cruisers which did not stipulate any control clause as it was
at the time judged impossible to build the ships secretly
(see Aron [1962]).

The next two lemmas demonstrate that, under reasonable
topological assumptions, deterring scenarios attached to
Pareto-optimal outcomes are feasible in all normal-form games.

Lemma 1.

*Notations are the same as those used in Definition 1.
Let $(x, \xi_{-i}; i = 1, \ldots, n)$ be a deterring scenario. Then x*

is an individually rational outcome

$$\sup_{y_i} \inf_{y_{-i}} u_i(y_i, \, y_{-i}) \le u_i(x) \tag{3}$$

for all $i \in N$. Conversely, suppose X_i is compact and u_i continuous, for all i. Then G possesses at least one individually rational outcome. For any such outcome x there exists for all $i \in N$ a threat ξ_{-i} against i such that $(x, \xi_{-i}; i = 1, \ldots, n)$ is a deterring scenario.

Proof.

From (2) we deduce

$$\inf_{y_{-i}} u_i(y_i, \, y_{-i}) \le u_i(y_i, \, \xi_{-i}(y_i)) \le u_i(x)$$

for all $y_i \in X_i$. Hence the first statement of Lemma 1. Conversely we know that under our topological assumption each player has at least one prudent strategy, say x_i (see Chapter 2); then, outcome $x = (x_i)_{i \in N}$ is individually rational. Next, for any i and $y_i \in X_i$, $y_i \ne x_i$, we pick an element $y_{-i} = \xi_{-i}(y_i) \in X_{-i}$ such that

$$u_i(y_i, \, y_{-i}) = \inf_{z_{-i}} u_i(y_i, \, z_{-i}) \le \sup_{z_i} \inf_{z_{-i}} u_i(z_i, \, z_{-i}) \le u_i(x)$$

QED.

Definition 2.

An imputation of game $G = (X_i, \, u_i; \, i = 1, \ldots, n)$ is a Pareto-optimal, individually rational outcome. We denote by $I(G)$ the set of imputations of G.

Lemma 2.

If X_i are compact and u_i continuous for all $i \in N$, the game G possesses at least one imputation.

Proof.

Denote by IR(G) the nonempty compact subset of individually rational outcomes of G. Next choose an element x of IR(G) that maximizes the function $\sum_{i \in N} u_i$ over IR(G). Then x is a Pareto-optimal outcome. Suppose, on the contrary, that y Pareto dominates x. Then y belongs to IR(G) and is such that $(\sum_{i \in N} u_i)(x) < (\sum_{i \in N} u_i)(y)$, a contradiction. QED.

The set I(G) of all imputations is the maximal range for cooperative negotiation. In general, it encompasses a wide range of payoff vectors. An exception would be the inessential games (Chapter 2) where all imputations have the same payoff $(\alpha_1, \ldots, \alpha_n)$. For instance, in two-player games the wolf-sheep lemma (Lemma 3, Chapter 2) picks the imputation most favorable to the wolf and exhibits the threat necessary to force the compliance of the sheep.

2. COALITIONAL THREATS: THE α CORE

We attack now the cooperative process in its full generality, where communication is possible within any sub-set of players and may result in some of these subsets coalescing. Any subset of players is thought of as a potential <u>coalition</u>, and coalitions form and disband for strategical reasons only. In this environment the notions of self-enforcing agreement and deterring threats become more complex. In this section we generalize deterring scenarios by allowing joint deviations by any coalition.

Definition 3.

Given a game $G = (X_i, u_i; i = 1, \ldots, n)$, the α core of G is the subset, denoted $C_\alpha(G)$ of those outcomes x^*, such that, for all coalition T, $T \subset N$ and all joint strategies $x_T \in X_T$, there exists a joint strategy $x_{T^c} \in X_{T^c}$ such that

$$u_i(x_T, x_{T^c}) \geq u_i(x^*) \qquad \text{for all } i \in T$$

<u>No</u>

$$u_i(x_T, x_{T^c}) > u_i(x^*) \qquad \text{for at least one } i \in T$$

This definition is best interpreted by means of a deterring scenario where coalitions react to joint deviations by the complement coalition. Formally, $(x^*, \xi_T,$ for all $T \subset N)$, where ξ_{T^c}, a mapping from X_T into X_{T^c}, is such that for no coalition $T \subseteq N$ and no joint strategy $x_T \in X_T$ we have

$$u_i(x_T, \xi_{T^c}(x_T)) \geq u_i(x^*) \quad \text{for all } i \in T$$

$$u_i(x_T, \xi_{T^c}(x_T)) > u_i(x^*) \quad \text{for some } i \in T$$

215

Then outcome x* belongs to $C_\alpha(G)$ iff there exists for all
$T \subseteq N$ a threat by coalition T^C against potential deviations
by coalition T such that $(x^*, \xi_T; T \subseteq N)$ is a coalitionally
deterring scenario. By definition 3, an outcome x* is in the
α core of G if every joint deviation x_T by coalition T can
be "countered" by a move x_{T^C} of the complement coalition T^C
that deters <u>at least one</u> member of T from entering deviation
x_T since he eventually becomes worse off: $u_i(x_T, x_{T^C}) < u_i(x^*)$
(or else every member of T keeps the same utility level after
all: $u_i(x_T, x_{T^C}) = u_i(x^*)$ for all i ε T).

<u>Lemma 3.</u>

  *The α core $C_\alpha(G)$ is a subset of $I(G)$, the set of imputations.
In a two-player game with compact strategy sets and continuous
utilities these two sets coincide: $C_\alpha(G) = I(G)$.*

<u>Proof.</u>
Taking T = N in Definition 3 shows that an outcome x* in
$C_\alpha(G)$ is Pareto-optimal. Taking T = {i} (the coalition made
up of Player i only) yields

$$\forall\, x_i \in X_i \; \exists\, x_{-i} \in X_{-i}: \; u_i(x_i, x_{-i}) \le u_i(x^*) \tag{4}$$

which implies $\sup_{x_i} \inf_{x_{-i}} u_i(x) \le u_i(x^*)$. Thus, x* is individually
rational. To prove the last statement, observe that, under
the topological assumptions property, (4) is equivalent to
$\sup_{x_i} \inf_{x_{-i}} u_i \le u_i(x^*)$. <u>QED.</u>

For two-player games, the α core is not empty (under the usual topological assumptions) since the imputation set is not. But for games with three players or more, the inclusion $C_\alpha(G) \subset I(G)$ is strict. In fact the α core $C_\alpha(G)$ might be empty even in finite games (of which the imputation set $I(G)$ is nonempty).

Example 2. The Condorcet paradox.

Let N be a society with an odd number of players, who must pick one among a finite set A of candidates. The voting rule is plurality voting. That is, each player casts a vote for one candidate and the candidate with the highest number of votes wins. In any normal-form game representing this voting rule (there are several such games since there are many ways to solve possible ties) and endowing each player with linear preferences over the candidates (indifferences ruled out), the α core is made up of those ballots that enforce election of a Condorcet winner, namely a candidate that defeats every other candidate in pairwise contests.

A typical situation where there is no Condorcet winner (and therefore the α core of the corresponding voting game is empty) occurs when $n = 3$, A contains three candidates, and

$$u_1(a) > u_1(b) > u_1(c)$$

$$u_2(b) > u_2(c) > u_2(a)$$

$$u_3(c) > u_3(a) > u_3(b)$$

When the α core of a game is empty, cooperative stability
cannot be achieved by deterring threats only. Even
strategic deviations that are apparent do not prevent
coalitions from making profitable moves that cannot be
countered by an appropriate reaction of the nondeviating
players. To restore stability, we invoke a more-subtle
behavioral scenario where the reaction by the nondeviating
players is to <u>bribe</u> some members of the betraying coalition
in such a way that the nonbribed betrayers eventually suffer
a loss.

Consider for instance the plurality-voting game with
three players and three candidates and in which the preferences
involve a Condorcet paradox as above. Choose an outcome
$x = (b, b, c)$, where b is the elected candidate. Stability of
x is jeopardized by coalition {1, 3}: by jointly voting for
a, Players 1 and 3 enjoy a strictly better utility level, not
threatened by any strategical reaction of Player 2:

$$u_i(a, x_2, a) > u_i(b, b, c) \qquad \text{for } i = 1, 3 \text{ and}$$
$$\text{all } x_2 \in X_2$$

However, Player 2 still can offer a better deal to Player 3.
By both voting for c, Players 2 and 3 force the election of
a, thus guaranteeing to Player 3 his optimal utility level;
at the same time, Player 1's utility falls to his worst level.
Anticipating that Player 2 will bribe Player 3, Player 1 is
deterred from entering the initially deviating coalition

{1, 3}, since he ends up below his initial utility level:
$u_1(c) < u_1(b)$.

The two-step scenario described above, where a coalitional move generates a countercoalition that bribes some players of the initial coalition eventually to deter the other deviating players, leads to a stability concept generalizing the α core notion. Although the technical complexity of the threat and counterthreat scenario is high, and its descriptive power is debatable, the corresponding equilibrium set proves always to be nonempty (Laffond-Moulin [1981]). Exercise 3 explores this result for n = 2.

Several analogous two-step stability concepts are available in the literature. See Rosenthal [1972] for a survey.

Another line of argument seeks to improve the stability property of the α core (hence looking for smaller equilibrium sets) by analyzing more finely the deterring scenarios, especially with respect to credibility and feasibility. Chapter 10 is devoted to this analysis.

3. GAMES IN CHARACTERISTIC FORM

Throughout this section, and this section only, we suppose that the cooperative output of any coalition can be measured by a numeraire, i.e., a transferable private good for which players can make side payments. This contrasts sharply with our systematically _ordinal_ viewpoint, where only individual preference orderings matter, with the exception of mixed strategies (see Chapter 7).

Definition 4.

   Given a finite society N, a characteristic-form game is a mapping v from the set P(N) of nonempty coalitions of players into R.

A game in characteristic form is a mathematical object which is dramatically simpler than a game in normal form. Much of the subtleties of strategic interaction disappear in this framework. Yet problems of cost allocations and/or sharing of joint benefits are enlightened by this formulation. Most of these "accounting" games share the following additional property.

Definition 5.

   The game (N, v) is said to be superadditive if v is such that, for all disjoint T, S ∈ P(N) we have

$$v(T) + v(S) \leq v(T \cup S)$$

Although superadditivity is not technically necessary for most of our results below, such as Theorem 1, it is needed to interpret the number $v(T)$ as the maximal joint profit resulting from cooperation within coalition T, independently of the complement coalition T^c. The inequality above expresses that coalition $T \cup S$ can do at least as well as by letting T and S cooperate independently.

Superadditivity of v clearly implies that, for all partition $S_1, \ldots, S_K$ of N

$$\sum_{k=1}^{K} v(S_k) \leq v(N)$$

Thus no partition $S_1, \ldots, S_K$ can yield a total joint profit above the profit $v(N)$ achieved by the grand coalition. Co-operation of all players is the only efficient cooperative outcome. The next two definitions parallel those for normal form games and bear the same name.

Definition 6.

An imputation *of the characteristic-form game v is a vector* $x = (x_i)_{i \in N}$ *of* R^N *such that*

$$\sum_{i \in N} x_i = v(N); \quad x_i \geq v(\{i\}) \qquad \text{for } i \in N$$

We denote by $I(v)$ *the set of imputations of game* (N, v).

Applying iteratedly superadditivity yields

$$\sum_{i \in N} v(\{i\}) \leq v(N).$$

Hence the set $I(v)$ of imputations is nonempty for superadditive games. Since $v(\{i\})$ is the secure profit level of which Player i is guaranteed by himself, it plays the role of the secure utility level of normal form games.

Definition 7.

The <u>core</u> of (N, v) *is the subset of imputations* $x \in I(G)$ *such that for all* $S \subseteq N$

$$\sum_{i \in S} x_i \geq v(S)$$

It is denoted $C(v)$.

Let our players bargain on the choice of a cooperative agreement. By superadditivity of v, the agreement requires the overall cooperation of the grand coalition N. The discussion bears on the division of the joint profit $v(N)$, i.e., the choice of a vector $x \in R^n$ such that $\sum_{i \in N} x_i = v(N)$. Individual rationality, $v(\{i\}) \leq x_i$, all $i \in N$, is a minimal requirement to obtain the consent of Player i. Therefore the players bargain on the choice of a particular imputation x. Against x any coalition S could form and ask for a better allocation (e.g., some $y \in I(v)$ such that $y_i > x_i$, for all $i \in S$), threatening to break down the overall cooperation (a very feasible threat since the unanimous consent of all players is needed to achieve the joint profit $v(N)$). To counter this threat, suppose that the players in S^c react by refusing once and for all to cooperate with any member of S. Then coalition S is left after all with a maximal joint profit $v(S)$, and condition $\sum_{i \in S} x_i \geq v(S)$ is the deterring property of that threat by S^c. Thus the core of (N, v) is the set of those divisions of $v(N)$ that are stabilized by the — natural —

coalitional threats. There is no longer any cooperation
with any coalition claiming for a better share of v(N).

Besides the interpretation of the core as resulting from
deterring threats described above, we have an equivalent
normative interpretation. Let x ε I(v) be an imputation of
game (N, v). Then x belongs to the core C(v) iff for all
S ⊂ C

$$\sum_{i \in S} x_i \leq v(N) - v(S^c)$$

Thus the core is made up of those imputations where no
coalition S receives more than its marginal contribution
$v(S \cup S^c) - v(S^c)$ to the profit of the grand coalition N.
(The obvious proof is omitted.)

Example 3. Cost sharing (Young et al [1982]).

Three township councils cooperate to build a water system
and must allocate its cost. The data are the cost figures for
all subsets of towns and the estimated benefit to each town.

Cost of a water system for

any town alone:	800
{1, 2} only:	1300
{2, 3} only:	1250
{1, 3} only:	1300
{1, 2, 3}:	1900

Benefits of any system to

town 1: 1000

town 2: 1100

town 3: 800

Hence the characteristic form

$$\begin{cases} v(1) = 1000 - 800 = 200 \\ v(2) = 1100 - 800 = 300 \\ v(3) = 800 - 800 = 0 \end{cases}$$

$$\begin{cases} v(12) = 1000 + 1100 - 1300 = 800 \\ v(23) = 1100 + 800 - 1250 = 650 \\ v(13) = 1000 + 800 - 1300 = 500 \\ v(123) = 1000 + 1100 + 800 - 1900 = 1000 \end{cases}$$

The core is given by the system of inequalities

$$x_1 \geq 200 \quad x_2 \geq 300 \quad x_3 \geq 0$$

$$x_1 + x_2 \geq 800 \quad x_2 + x_3 \geq 650 \quad x_1 + x_3 \geq 500$$

$$x_1 + x_2 + x_3 = 1000$$

A better intuition of the core follows by normalizing our game in such a way that the imputation set is just the unit simplex of R^3:

$$w(i) = 0, \quad w(123) = 1, \quad w(ij) = \frac{v(ij) - v(i) - v(j)}{v(123) - v(1) - v(2) - v(3)}$$

Now the core is given by

$$x_i \geq 0 \quad \text{for all } i = 1, 2, 3; \quad x_1 + x_2 + x_3 = 1$$

$$x_1 + x_2 \geq .6; \quad x_2 + x_3 \geq .7; \quad x_1 + x_3 \geq .6$$

It is easily seen to be a triangle, the convex hull of three imputations

$$(.3, .3, .4), (.3, .4, .3) \text{ and } (.2, .4, .4)$$

Returning to the original game, its core is the convex hull of the following three imputations.

$$(350, 450, 200) \quad (350, 500, 150) \quad (300, 500, 200)$$

Thus all players' payoffs are determined with an error of at most \$25. A typical member of the core is the center of $C(v)$:

$$x^* = (333.3, 483.3, 183.3)$$

At x^*, each two-player coalition makes the same extra profit $x_i + x_j - v(\{i, j\}) = 16.6\$$. Imputation x^* is a fair compromise within $C(v)$. Along this line of argument, a normative arbitration concept, the nucleolus (Schmeidler [1969]) can be defined for any game.

Emptyness of the core, just like that of the α core in normal-form games, occurs when intermediate coalitions are too powerful. This is particularly clear in our next example.

Example 4. Symmetrical games.

A symmetrical game gives identical power to coalitions with the same size. The characteristic form v is such that

$$v(S) = v^*(s)$$

where $s = |S|$, for all $S \subseteq N$. The imputation set of V* is simply the following simplex of R^n:

$$x \in I(v^*) \iff \sum_{i=1}^{n} x_i = v^*(n), \ x_i \geq v^*(1), \ i = 1, \ldots, n$$

The core is defined by a symmetrical system of linear inequalities. Therefore it is nonempty iff it contains the center x* of I(v*), namely, $x_i^* = (v^*(n))/n$. This amounts to

$$\frac{1}{s}v^*(s) \leq \frac{1}{n}v^*(n)$$

for all $s = 1, 2, \ldots, n$.

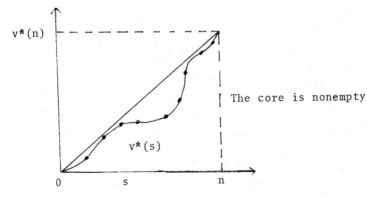

The core is nonempty

Figure 1

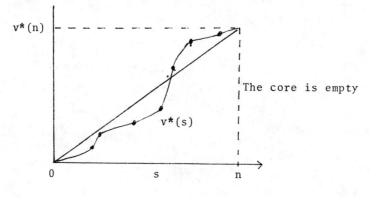

Figure 2

As an application, consider the following treasure-hunt
game due to Shapley. n men discover an unlimited hoard of
gold ingots in the mountains. It takes two men to carry out
a single ingot (while protecting themselves!) and there is
no opportunity to return to the mountains for another load.
The characteristic function is

$$v(t) = \frac{t}{2} \qquad \text{if t is even}$$

$$\frac{t - 1}{2} \qquad \text{if t is odd}$$

The core is the singleton x_i = 1/2, i = 1, ..., n if n is
even. If n is odd, the core is empty. Returning to the
general case, we characterize now the nonemptyness of the core
with the help of linear programming.

227

Definition 8.

Given N and a Player _i_ we denote by P(_i_) the set of nonempty coalitions containing player _i_:

$$\{S \in P(i) \iff \{S \in P(N) \text{ and } i \in S\}$$

A balanced family of coalitions is a mapping δ from P(N) into [0, 1] such that for all _i_ ∈ N

$$\sum_{S \in P(i)} \delta_S = 1 \qquad (5)$$

Finally, we say that a game in characteristic form (N, v) is __balanced__ if it satisfies

$$\sum_{S \in P(N)} \delta_S v(S) \leq v(N) \qquad (6)$$

for all balanced family of coalitions δ.

Interpreting the notion of a balanced family of coalitions is not all that easy. Examples are any partition of N, or the set of coalitions with the same fixed size. For n = 3, 4, it is not too difficult to find the set of minimal balanced family of coalitions, corresponding to the extreme points of the convex polyhedron in $[0,1]^{P(N)}$ defined by (5). We have __n = 3__: all partitions, e.g., {1} {23}, δ(1) = δ(23) = 1, δ(S) = 0 otherwise. {12} {23} {13}, δ(ij) = 1/2, δ(S) = 0 otherwise. __n = 4__: all partitions. {12} {23} {34} {41}: δ = 1/2 on those four, δ = 0 otherwise. All 3-players coalitions: $\delta = 1/3$ on those, $\delta = 0$ otherwise. $\{123\}$ $\{234\}$ $\{14\}$: $\delta = 1/2$ on those three, $\delta = 0$ otherwise. $\{12\}$ $\{23\}$ $\{31\}$ $\{4\}$: $\delta = 1/2$ on the first three, $\delta(4) = 1$, $\delta = 0$ otherwise.

Theorem 1 (Bondareva [1962], Scarf [1967]).

The game (N, v) has a nonempty core iff it is balanced.

Proof.

If x belongs to $C(v)$ and δ is a balanced family of coalitions, we have

$$x(S) \geq v(S) \implies \delta_S x(S) \geq \delta_S v(S)$$

for all $S \subset N$. By summing up these inequalities we get

$$\sum_{S \in P(N)} \delta_S v(S) \leq \sum_{S \in P(N)} \delta_S x(S) = \sum_{i \in N} \sum_{S \in P(i)} \delta_S x_i = \sum_{i \in N} x_i = v(N)$$

Conversely, if the core $C(v)$ is empty, the hyperplane $\sum_{i \in N} x_i = v(N)$ is disjoint from the convex (nonempty) polyhedron:

$$\text{for all } S \subseteq N \quad \sum_{i \in S} x_i \geq v(S)$$

By a standard separation argument, this implies the existence for all S of a nonnegative number δ_S such that

$$
\begin{cases}
\text{for all } x \varepsilon R^N: \quad \sum_{i \in N} x_i = \sum_{S \in P(N)} \delta_S (\sum_{i \in S} x_i) \\
\text{and} \\
\sum_{S \in P(N)} \delta_S v(S) > v(N)
\end{cases}
$$

The first of these properties is equivalent to saying that δ is a balanced family of coalitions. QED.

Corollary.

A *three-player game has a nonempty core iff*

$$\begin{cases} v(1) + v(23) \leq v(123) \\ and \\ two \text{ } similar \text{ } inequalities \text{ } obtained \text{ } by \text{ } permuting \text{ } players \\ and \\ v(12) + v(23) + v(13) \leq 2 \text{ } v(123). \end{cases}$$

A *four-player, superadditive game has a nonempty core iff*

$$\begin{cases} v(12) + v(23) + v(34) + v(41) \leq 2 \text{ } v(1234) \\ v(123) + v(234) + v(134) + v(124) \leq 3 \text{ } v(1234) \\ v(123) + v(234) + v(14) \leq 2 \text{ } v(1234) \\ and \\ seven \text{ } similar \text{ } inequalities \text{ } derived \text{ } by \text{ } permuting \text{ } the \text{ } players. \end{cases} \qquad (7)$$

Proof.

If δ is associated to a partition of N, inequality (6) follows from superadditivity of v. If δ is the minimal balanced family {12} {23} {31} {4}, inequality (6) follows again from superadditivity of v and

$$v(12) + v(234) + v(134) \leq 2 \text{ } v(1234)$$

since

$$v(23) + v(4) \leq v(234), \text{ } v(13) + v(4) \leq v(134)$$

Thus the ten inequalities (7) are sufficient, given that v is superadditive. <u>QED</u>.

Emptyness of the core does not imply that cooperation of the overall coalition N is impossible. It simply means that no imputation can be stabilized by the simple, natural threats described above. In this case cooperative stability requires a more complex behavioral scenario similar to the threat and counter-threats described in Section 2. We give but one example of these two-stage concepts. Many more have been developed, among them the von Neumann and Morgenstern solution and the bargaining set (Aumann and Maschler [1964]). None of those, however, is simple and tractable enough to share the wide applicability of the core concept. They will survive, hopefully, as fascinating creatures from the game-theoretical folklore.

<u>Example 5.</u> <u>Objections and counterobjections in a three-player game with an empty core.</u>

Let v be a superadditive, three-player game with an empty core. Without loss of generality, we assume

$$v(i) = 0, \; i = 1, 2, 3, \; v(123) = 1$$

$$v(12) = a_3, \; v(23) = a_1, \; v(13) = a_2, \; 0 \le a_i \le 1$$

Suppose v has an empty core $a_1 + a_2 + a_3 > 2$. Two types of imputations x can be stabilized.

<u>Type 1</u>: $x_i + x_j < a_k$: each two-player coalition has an objection against x. For instance, an objection of {1, 2} is a pair (y_1, y_2), such that $x_i < y_i$, $i = 1$, 2, and $y_1 + y_2 = a_3$. However, Player 3 has a plausible counter-objection if he can bribe either Player 1 or Player 2 and restore his (Player 3's) original payoff:

$$\exists z_2, z_3 \quad \begin{matrix} z_2 > y_2 \\ z_3 \geq x_3 \\ z_2 + z_3 = a_1 \end{matrix} \quad \text{and/or} \quad \exists z_1, z_3 \quad \begin{matrix} z_1 > y_1 \\ z_3 \geq x_3 \\ z_1 + z_3 = a_2 \end{matrix}$$

which is equivalent to

$$y_2 < a_1 - x_3 \quad \text{and/or} \quad y_1 < a_2 - x_3$$

Thus, any objection of {1, 2} can be countered by 3 via a counterobjection by {2, 3} and/or {1, 3} iff

$$a_3 < (a_1 - x_3) + (a_2 - x_3) \iff x_3 < \frac{1}{2}(a_1 + a_2 - a_3)$$

Hence, the subset D_1 of "stable" imputations:

$$x_1 < \frac{1}{2}(a_2 + a_3 - a_1)$$

$$x_2 < \frac{1}{2}(a_1 + a_3 - a_2)$$

$$x_3 < \frac{1}{2}(a_1 + a_2 - a_3)$$

<u>Type 2</u>: $x_1 + x_2 \geq a_3$, $x_2 + x_3 < a_1$, $x_1 + x_3 < a_2$. Players 1 and 2 successfully collude.

Say that y_2, y_3 is an objection of $\{2, 3\}$:

$$x_i < y_i \quad i = 2, 3, \quad y_2 + y_3 = a_1$$

Player 1 can offer a better deal to Player 3 if there exists z_1, z_3 such that

$$y_3 < z_3, \quad x_1 \leq z_1 \quad \text{and} \quad z_1 + z_3 = a_2$$

which is possible iff $y_3 < a_2 - x_1$. As the upper bound of y_3 in an objection with Player 2 is $a_1 - x_2$, we conclude that Player 1 can counter any unfaithfulness by Player 2 iff $a_1 - x_2 \leq a_2 - x_1$. A symmetrical argument shows that equality is necessary to enforce stability of the collusion $\{1, 2\}$. Hence the subset D_2 of stable imputations:

$\{x_3 \geq 0, \; x_1 + x_2 \geq a_3, \; x_1 + a_1 = x_2 + a_2\}$: 1 and 2 collude

or

$\{x_1 \geq 0, \; x_2 + x_3 \geq a_1, \; x_2 + a_2 = x_3 + a_3\}$: 2 and 3 collude

or

$\{x_2 \geq 0, \; x_1 + x_3 \geq a_2, \; x_1 + a_1 = x_3 + a_3\}$: 1 and 3 collude

These subsets are depicted on Figure 3. The large triangle is the imputation set. The shaded triangle is D_1. The three thick segments form D_2.

Similar scenarios where two players successfully collude may even threaten the stability of the core. For instance,

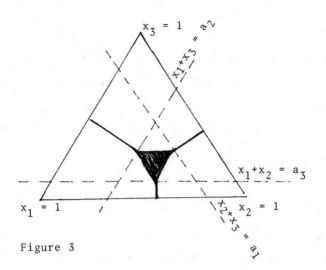

Figure 3

suppose that for any element x ε C(v) we have

$$x_1 + x_2 < a_3 + \frac{1 - a_3}{2} = \frac{1}{2} + \frac{a_3}{2} \tag{8}$$

Notice that we still work with a normalized game where $v(i) = 0$, $v(123) = 1$, $v(ij) = a_k$ as in the above example. Inequality (8) means that Players 1, 2 will do better than anywhere in the core if i) they collude, and ii) as a single player bargaining against Player 3 they obtain 1/2 of the cooperative surplus in the resulting two-player game (namely, $v(\{12\}) = a_3$, $v(3) = 0$, $v(\{12\}, 3) = 1$). Since $\sup_{C(v)} (x_1 + x_2) = \min \{1, 2 - a_1 - a_2\}$ (as the reader can check easily), the situation x ε C(v) implies (8) corresponds to $a_1 + a_2 + (a_3/2) > (3/2)$. This is compatible with nonemptyness of the core $(a_1 + a_2 + a_3 < 2)$ and even with a symmetrical game:

$(3/5) < v(ij) = v^*(2) \leq (2/3)$.

After these rather elaborate arguments for three-player games, hopefully the reader is convinced that the analysis of four-player games is virtually intractable. This intuition is confirmed by more than three decades of theoretical research.

4. EXERCISES

a) Imputations and α core

1) A three-player, zero-sum game

Three players 1, 2, 3 rotate taking one or two sticks from a pile originally containing $n \geq 1$ sticks. Player 1 goes first followed by Player 2, then by Player 3, then by Player 1 again (if the pile is not exhausted) and so on. Say that i is the player who takes the last stick; then, the payoff is

$$-3 \text{ to Player } i$$
$$+2 \text{ to Player } i + 1$$
$$+1 \text{ to Player } i + 2$$

(where we set $3 + 1 = 1$, $3 + 2 = 2$, $2 + 2 = 1$).

a) Compute the perfect equilibrium of this game for all n.
b) Prove that coalition {12} can force 3 to lose except for $n = 1, 2, 6$, coalition {23} can force 1 to lose except for $n = 2, 3, 4, 7, 8$, coalition {13} can force 2 to lose except for $n = 1, 4, 5, 9$.   c) Deduce that for $n \geq 10$ the α core of the game is empty. Compute it for $3 \leq n \leq 9$.

Exercises 2 and 3 are more difficult, and should be skipped by the nonabstract reader.

2) Let G by a symmetrical game. For any permutation σ of $\{1, \ldots, n\}$ and any strategy n-tuple $x \in X_N$, denote x^σ the following strategy n-tuple:

$$(x^\sigma)_i = x_{\sigma(i)}$$

for all $i = 1, \ldots, n$. Then we have

$$u_{\sigma(i)}(x^\sigma) = u_i(x)$$

for all $i = 1, \ldots, n$, and all $x \in X_N$.

Suppose G has compact strategy sets and continuous payoff functions. Prove that the α core of G is nonempty.

3) <u>Objections and counterobjections in three-player games</u>
    (Laffond - Moulin [1981])

Let $(X_1, X_2, X_3; u_1, u_2, u_3)$ be a three-player game with <u>finite</u> strategy sets X_i. Given an outcome $x \in X_{\{1, 2, 3\}}$ we denote by $0_{12}(x)$ the set of objections by coalition $\{1, 2\}$ against x:

$$[(y_1, y_2) \in 0_{12}(x)] \iff [\inf_{y_3} u_i(y_1, y_2, y_3) > u_i(x_1, x_2, x_3)$$

for $i = 1, 2]$.

A counterobjection by coalition $\{2, 3\}$ against objection

236

$(y_1, y_2) \in 0_{12}(x)$ is a pair $(z_2, z_3) \in X_2 \times X_3$ such that

$$\inf_{z_1} u_2(z_1, z_2, z_3) > \inf_{y_3} u_2(y_1, y_2, y_3)$$

$$\sup_{z_1} u_1(z_1, z_2, z_3) < u_1(x_1, x_2, x_3)$$

(9)

Player 3 forces the cooperation of Player 2: If you stick to objection (y_1, y_2), I will keep your utility level not above $\inf_{y_3} u_2(y_1, y_2, y_3)$, whereas by agreeing to join me in (z_2, z_3) you are guaranteed of the utility level $\inf_{z_1} u_2(z_1, z_2, z_3)$. Finally, the bottom inequality in (9) deters Player 1 from entering objection (y_1, y_2). The <u>deterrence set</u> of G is made up of all imputations such that to any objection by {i, j} there exists a counterobjection by {i, k} and/or by {j, k}.

Prove that the deterrence set of G is nonempty.
<u>Hint</u>: Assume first $I(G) = X_{\{123\}}$. Say that an objection $(y_1, y_2) \in 0_{12}(x)$, by coalition {12} against x, is maximal if i) it has no counterobjection, and ii) for all $(z_1, z_2) \in 0_{12}(x)$: $\{\inf_{z_3} u_i(z_1, z_2, z_3) \geq \inf_{y_3} u_i(y_1, y_2, y_3)$ for i = 1, 2$\} \Rightarrow$ {these inequalities are equalities}.

Next, suppose that the deterrence set of G is empty and construct inductively a sequence

$$x^0, (S^1, x^1), (S^2, x^2), \ldots, (S^t, x^t), \ldots$$

where x^o is arbitrary in $X_{\{123\}}$, and x^t is a maximal objection of coalition S^t against x^{t-1} with no counterobjection (for all $t \geq 1$). Then prove that any two outcomes x^t differ and derive a contradiction.

b) <u>Games in characteristic form</u>

4) A symmetrical n-person game has $v(S) = -1$ for all coalitions S. a) Prove the core is nonempty. b) Define it by n + 1 inequalities instead of $2^n - 1$.

<u>Hint</u>: Start with n = 3, 4 and generalize.

c) Describe its most unfair elements. (By "unfair" we mean that the difference between the richest and poorest players is as large as possible).

5) <u>Pollute the lake</u>

There are n factories around a lake. It costs B (per day) for a factory to treat its wastes before discharging them into the lake. It costs kA (per day) for a factory to purify its water supply, if k is the number of factories that <u>do</u> <u>not</u> treat their wastes. Assume $A \leq B \leq nA$. a) Determine the characteristic function. b) When does this game have a core? Does it every have a 1-point core? If so, when?

6) <u>An interesting core</u>

Assume that n is <u>even</u> and consider the n-person game with $v(N) = n/2$.

$$v(1,\ 2) = v(3,\ 4) = \ldots = v(n-3,\ n-2) = v(n-1,\ n) = 1$$

$$v(2,\ 3) = v(4,\ 5) = \ldots = v(n-2,\ n-1) = v(n,\ 1) = 1$$

and $v(S)$ is the minimum determined by superadditivity for all other $S \subset N$. Show that the core is the line segment between the "odd" imputation $(1,\ 0,\ 1,\ 0,\ \ldots,\ 1,\ 0)$ and the "even" imputation $(0,\ 1,\ 0,\ 1,\ \ldots,\ 0,\ 1)$.

7) **Inessential games**

Say that the characteristic function v is inessential if its imputation set is a singleton:

$$v(N) = \sum_{i \in N} v(i)$$

Show that a game is {superadditive and inessential} iff it is {additive}:

$$\text{for all coalition } S: \quad v(S) = \sum_{i \in S} v(i)$$

8) **Convex games** (Shapley [1971])

Say that $(N,\ v)$ is a convex game if

$$v(S) + v(T) \le v(S \cap T) + v(S \cup T)$$

for all $S,\ T \subset N$.

a) Prove in this case

$$v(N) - v(S^C) = \sup_{T \subseteq S^C} \{v(S \cup T) - v(T)\}$$

for all $S \subset N$.

b) For any ordering 1, 2, ..., n of N, set

$$x_i = v(1, 2, ..., i) - v(1, 2, ..., i - 1), i = 2, ..., n$$

$$x_1 = v(1)$$

Prove that x is in the core.

c) In fact, $C(v)$ is the convex hull of the imputations x_σ obtained as in b) when the ordering σ of N varies. Prove this for n = 3.

9) Simple games

We say that game (N, v) is <u>simple</u> if

$$\text{for all } S \subseteq N, v(S) = 0 \text{ or } 1$$

In that case we denote by W the set of winning coalitions, namely:

$$S \in W \iff v(S) = 1$$

a) Prove that the simple game (N, v) is superadditive iff W is monotonic and proper:

monotonic: $S \in W, S \subseteq T \implies T \in W$ for all S, T

proper: $S \in W \implies S^c \notin W$

From now on, assume that W is superadditive. b) Say that Player $i^* \in N$ is a dictator of W if $\{i^*\}$ is a winning coalition. Assuming that (N, v) is superadditive, prove that W has a (unique) dictator iff it is inessential. c) Suppose that W

has no dictator and is nonempty. Say that Player i* is a
veto player if N\{i*} is not winning. Denote by N_* the
possibly empty set of veto players. Prove that the core
$C(v)$ is nonempty iff N_* is nonempty. In this case $x \in C(v)$
iff

$$\sum_{i \in N_*} x_i = 1, \quad x_i = 0 \quad \text{for all } i \notin N_*$$

$$x_i \geq 0 \quad \text{for all } i \in N_*$$

d) Let for all i, Player i be endowed with a voting weight
$q_i \geq 0$ and denote by $q_o \geq 0$ a quota such that

$$q_o \leq \sum_{i \in N} q_i < 2q_o$$

Then define the weighted majority game W_q by

$$S \in W_q = \sum_{i \in S} q_i \geq q_o$$

Prove that W_q is a monotonic and proper simple game. Prove
that i* is a dictator iff $q_o \leq q_{i*}$ and i* is a veto player iff

$$\sum_{i \in N} q_i - q_o < q_{i*}$$

CHAPTER 10. THREATS AND REPETITION

1. PASSIVE THREATS: STRONG EQUILIBRIUM

We explore several refinements of the α core (of normal-form games) based upon a finer analysis of the deterring scenarios.

Among deterring threats some are distinctly simpler. These include the passive threats suggesting that you will not react to any more by any subset of your fellow players. Loyalty to the agreement means that you stick to your agreed-upon strategy no matter what the rest of the world does or says. The only difference with the Nash-equilibrium scenario is the eligibility of joint deviation of a coalition of players.

Definition 1 (Aumann [1959]).

Given a N-normal-form game $G = (X_i, u_i; i = 1, \ldots, n)$, *we say that* x* *is a strong equilibrium outcome if no coalition of players can profitably deviate, given that the complement coalition will not react.*

$$\forall\, T \subset N \ \forall\, x_T \in X_T \quad No \left\{ \begin{array}{l} \forall\, i \in T \ \ u_i(x_T, x^*_{T^c}) \geq u_i(x^*) \\[2ex] \exists\, i \in T \ \ u_i(x_T, x^*_{T^c}) > u_i(x^*) \end{array} \right.$$

We denote by SE(G) *the — possibly empty — set of strong equilibrium outcomes of G.*

Taking T = N yields that a strong equilibrium is a Pareto-optimum. Taking T = {i}, i ∈ N, yields that it is a Nash equilibrium as well. Clearly in two-person games a strong equilibrium simply is a Pareto-optimal Nash equilibrium.

Also, a strong equilibrium is in the α core: SE(G) ⊆ C_α(G) with the associated passive threats $\xi_{T^c}(x_T) = x^*_{T^c}$ for all T.

An alternative interpretation of the stability property in Definition 1 goes by a two-stage decision-making process. In the first stage our players congregate to agree upon a particular outcome x*. Next communication is banned and each individual has to make a single isolated decision on his final irrevocable strategy. Each one can freely betray his word to play x^*_i but he cannot inform the nonbetraying players of his own deviation. Also, any coalition T can plot a jointly decided betrayal x_T, but again the players outside the conspiracy cannot be informed of the intended change of strategy. Therefore, they should be expected to stick to the agreed-upon behavior $x^*_{T^c}$. This limitation to communication is crucial to the plausibility of Definition 1 as our Example 2 below illustrates "a contrario." Notice that we do not invoke any tatonnement process to enhance the strong equilibrium concept, since the myopic behavior cannot be justified in a cooperative world.

Example 1.   Local public goods with congestion.

Each player must pick one among p local public goods. If t_k is the fraction of players deciding for the good k, k = 1, ..., p, then every consumer of k enjoys the utility level $-a_k(t_k)$ where $a_k(\cdot)$, the disutility of consuming good k, is a strictly increasing function of t_k. We have the following game.

$$X_i = \{1, \ldots, p\}$$

for all i ϵ N.

$$u_i(x) = -a_k(t_k)$$

if $x_i = k$ and $t_k = |\{j \epsilon N/x_j = k\}|/n$.

Assume that for all k = 1, ..., n, a_k increases continuously from $a_k(0) = 0$ to $a_k(1) = 1$. Then there is a unique vector $t^* = (t^*_1, \ldots, t^*_n)$ in the simplex ($t^*_i \geq 0 \sum_{i=1}^{n} t^*_i = 1$) such that $a_k(t^*_k) = a_\ell(t^*_\ell)$ all $k, \ell = 1, \ldots, p$ (indeed the inverse function a_k^{-1} is strictly increasing, for all k, so there is a unique λ such that $\sum_{i=1}^{n} a_k^{-1}(\lambda) = 1$).

Any outcome x* with the associated vector of fractions t* is a strong equilibrium. Namely, after coalition T deviates to x_T, the new fractions $t_1, \ldots, t_p$ are such that

$$\text{for at least one } i \in T, x_i \in t_k \text{ and } t_k > t^*_k$$

or

$$\text{for all } k = 1, \ldots, p, t_k = t^*_k$$

Thus coalition T cannot profitably deviate. For example,

$$\{x_i \in t_k \text{ and } t_k > t^*_k\} \Rightarrow \{u_i(x) < a_k(t^*_k) = u_i(x^*)\}$$

Assuming n is large makes the converse true as well. Any strong equilibrium of the game has corresponding fractions close to t*.

Example 2. <u>A bargaining game.</u>

The n players must share a dollar. They submit individual claims. These claims are satisfied if they are feasible altogether. Otherwise each player gets nothing.

$$X_i = [0, 1] \text{ for all } i \in N$$

$$u_i(x) = x_i \text{ if } \sum_{j \in N} x_j \leq 1$$

$$= 0 \text{ if } \sum_{j \in N} x_j > 1$$

We assert that

$$SE(G) = \{x \in X_N \mid \sum_{i \in N} x_i = 1\}$$

In other words, an outcome is a strong equilibrium iff the corresponding claims add up to exactly one dollar. We let the reader check this assertion.

Observe that if coalition T, acting as a leader, commits itself to a particular strategy T-tuple x^*_T with $\sum_{i \in T} x^*_i = 1 - \varepsilon$ ($\varepsilon > 0$ is small), then coalition T^c, acting as a follower, optimally replies by a vector of claims x_{T^c} such that

$$\sum_{i \in T^c} x_i = \varepsilon$$

Hence, any individual or coalition endowed with enough commitment power will nearly extract all the potential benefit of one dollar. The competition for the leadership is very intense in this game. Pick now a particular SE outcome x* upon which the players may have agreed, say $x^*_i = 1/n$, for all $i \in N$. To self-enforce this agreement, each player must commit himself never to listen to any claim above $1/n$ by anyone threatening to take the leadership. This policy is optimally achieved by a careful destruction of the communication channels among players, after which every player can still change at will his or her strategy but cannot do it publicly anymore.

In a game like Example 2, the crucial stability feature of the agreement to play a particular strong equilibrium relies on the restriction of communication that the players must enforce legally (as in a secret ballot regulation) or physically (like Ulysses' companions stopping their ears with wax).

In a normal-form game, the strategy sets of which are open subsets of euclidian spaces, and utility functions are differentiable, one would not expect the existence of a Pareto-optimal Nash equilibrium. A fortiori, it should not possess an SE outcome. In that limited sense existence of a strong equilibrium is exceptional. Under very stringent convexity assumptions Ichi-Ishi [1983] gives an existence result for strong equilibrium.

2. REPEATED GAMES AND THE β CORE

A deterring scenario (Definition 1, Chapter 8) relies on multilateral threats. Tit for tat succeeds in stabilizing the peaceful outcome because each player threatens his partner. To implement your threat, you must act as a follower. However, there is no way both players can be followers! To overcome this difficulty, we imagine that the game is played repeatedly and the overall payoff results from a suitable summation of the instantaneous payoffs. Presumably, if the short-run profit of a noncooperative deviation is outweighed by the long-run losses implied by tit for tat, then the deterring scenario succeeds in enforcing the peaceful cooperative outcome.

This statement is fairly intuitive, yet its technical realization is quite complicated. The starting point is to think of the repetition of the game as an <u>extension</u> of the initial game (very much like the mixed extension of Chapter 7) that endows each player with more strategies (yet the original strategies are still available) and defines an extended payoff function accordingly. For instance, if the initial game $G = (X_1, X_2, u_1, u_2)$ is repeated <u>twice</u> the extended game is $\tilde{G} = (X_1 \times X_1^{X_2}, X_2 \times X_2^{X_1}, \tilde{u}_1, \tilde{u}_2)$, where Player i's strategies (x_i, ξ_i) means that he plays x_i in the first period and $\xi_i(x_j)$ in the second (if the other player used x_j in the first period). The extended payoff is then

$$\tilde{u}_i((x_1, \xi_1), (x_2, \xi_2)) = \frac{1}{2}[u_1(x_1, x_2) + u_1(\xi_1(x_2), \xi_2(x_1))] \quad (1)$$

This repetition process, however short, generates new Nash-equilibrium outcomes based upon deterring threats.

Example 3. <u>A Cournot's quantity setting duopoly</u> (continued).

The game is that of Example 2, Chapter 6. The unique Nash equilibrium is $((1/6)/(1/6))$ with corresponding payoffs 1/36 to each player. The symmetrical Pareto-optimal outcome is $((1/8)/(1/8))$ with payoff 1/32 to each. This is the fair cooperative outcome where joint profit is maximized; each refrains from using his best reply $r_i(1/8) = (3/16)$, that would yield a payoff

$$u_i(\underset{i}{\tfrac{3}{16}} , \underset{j}{\tfrac{1}{8}}) = \frac{9}{16^2} > \frac{1}{32}$$

To enforce the cooperative behavior in the <u>first</u> period
of the twice-repeated game, each player threatens to flood
the market in the second period if his partner did not cooperate
in the first one. If he does, then he uses his Nash-equilibrium
strategy:

$$\bar{\xi}(x_j) = \frac{1}{2} \qquad\qquad \text{if } x_j \neq \frac{1}{8};$$

$$\bar{\xi}(\frac{1}{8}) = \frac{1}{6}$$

We claim that $((1/8), \bar{\xi})$, $((1/8), \bar{\xi})$ is a Nash equilibrium
in the extended game (1). Indeed, if Player 2 uses $((1/8), \bar{\xi})$,
Player 1 cannot gain by deviating in the second period only
because Player 2 uses a Nash-equilibrium strategy in that
period. He cannot gain by deviating in the first period
either because the gain $(9/16^2)$ - $(1/32)$ in that period is
smaller than the implied loss $(1/36)$ - 0 in the second period
(where Player 2 uses $x_2 = 1/2$ and prevents Player 1 from
making any profit.)

In the T-periods repetition of the game, a Nash equilibrium
obtains where both players act cooperatively in all periods
but the last one, while they are threatening to flood the
market forever as soon as the other player deviates once. The
corresponding payoff is nearly 1/32 to each.

A peculiar feature of finitely repeated games is that the
last period must be noncooperative. No threat is meaningful
when there is no tomorrow. In certain games, notably the

prisoner's dilemma, this rules out any "good" outcome from the Nash equilibrium of the repeated game. Knowing that in the last period both players are going to be aggressive anyway, no threat is a deterrent in the next to the last period. As a result, they both act aggressively in that period also, and so on. To overcome this difficulty it is necessary that the game should never end, which means that it is repeated forever. For a systematical treatment of repeated games with a finite horizon see Benoit and Krishna [1984].

For a repeated game with infinite horizon and discounted payoff we start with $G = (X_i, u_i; i = 1, \ldots, n)$ where the u_i are uniformly bounded on X_N (e.g., X_i are compact and u_i continuous, for all $i \in N$). We imagine that G is played once at $t = 1$, and let x^1 be the corresponding outcome. Next a random device commands the player to stop with probability $(1 - \delta)$ in which case the overall utility to Player i is $u_i(x^1)$, for all $i \in N$, or (with probability δ) to continue, in which case G is played again at time 2. And so on. After each play of G, it is decided with probability $(1 - \delta)$ that the corresponding outcome is the final outcome or with probability δ, that the entire history of plays is neglected and another play of G occurs.

Another, formally equivalent, story goes by saying that G is played infinitely many times anyway, and the overall

payoff resulting from the sequence of plays x^1, x^2, ...,
x^t, ... to Player i is

$$(1 - \delta) \{u_i(x^1) + \delta u_i(x^2) + \ldots + \delta^{t-1} u_i(x^t) + \ldots\} \qquad (2)$$

for all i ε N. We set $\beta_i = \inf_{x_{-i}} \sup_{x_i} u_i$ to be the maximal secure
utility level of Player i, i.e., the minimal payoff to
Player i when he can observe beforehand the strategies of all
other players. Now pick an imputation x* such that

$$\beta_i \leq u_i(x^*)$$

for all i ε N. We interpret x* as an agreement that the
players wish to enforce by discounted repetition. For that
purpose they will seek a Nash equilibrium σ* of the repeated
game which yields exactly payoff $u_i(x^*)$ fo Player i, for all
i ε N. Here σ^*_i, a strategy of Player i in the repeated
game, is complicated. It specifies for each past history
x^1, ..., x^{t-1} the strategy x^t_i that Player i uses at time t
(for all t ε ℕ). Actually, to enforce $u_N(x^*)$ as a Nash-
equilibrium payoff vector a very simple outcome σ* is in order.
At σ^*_i Player i uses the agreed-upon pure strategy x^*_i as long
as no Player j deviates from x^*_j. As soon as a deviating
Player j is detected, strategy σ^*_i commands that he be punished
forever.

   To describe formally our Nash equilibrium, we pick for
all j ε N a strategy N\{j}-tuple $\tilde{x}^j_{-j}$ such that

$$\sup_{x_j} u_j(x_j, \tilde{x}^j_{-j}) = \beta_j$$

For all $i \in N$ a strategy σ^*_i of Player i in the repeated game is chosen such that

$$x^1_i = x^*_i$$

At time 1, play x^*_i.

If $\qquad\qquad x^1 = x^2 = \ldots = x^{t-1} = x^*$

then $x^t_i = x^*_i$. At time t, play the agreed-upon outcome if everybody did the same so far.

If $\qquad\qquad x^1 = x^2 = \ldots = x^{t-2} = x^* \neq x^{t-1}$,

then pick a j such that $x^{t-1}_j \neq x^*_j$. Next play $\tilde{x}^j_i = x^t_i = x^{t+1}_i = \ldots$.

Under what conditions is it the case that the behavior depicted by σ^* deters any isolated individual from deviating? It has to be that the short-run profit to Player i of a deviation at time t (that cannot be countered before the next play) is overweighted by the long-run loss following the punishment by the players in $N\backslash(i)$ (who, by assumption stick to strategy $\sigma^*_{N\backslash\{i\}}$ at time $t + 1$, $t + 2$, ...). Setting

$$u^*_i(x^*_{-i}) = \sup_{x_i} u_i(x_i, x^*_{-i})$$

and comparing the discounted value at time t of the short-run profit $u_i^*(x_{-i}^*) - u_i(x^*)$ and the long-run losses, the self-enforcement property is then

$$u_i^*(x_{-i}^*) - u_i(x^*) \le \delta(u_i(x^*) - \beta_i) + \delta^2(u_i(x^*) - \beta_i) + \ldots$$

which is equivalent to

$$1 - \delta \le \frac{u_i(x^*) - \beta_i}{u_i^*(x_{-i}^*) - \beta_i}$$

for all $i \in N$.

When δ goes to one, this inequality is automatically satisfied. This suggests to consider the infinitely repeated game with a Cesaro mean payoff:

$$\lim_{\delta \to 1} (1 - \delta)\{u_i(x^1) + \ldots + \delta^{t-1}u_i(x^t) + \ldots\} =$$

$$\lim_{T \to +\infty} \frac{1}{T}\{u_i(x^1) + \ldots + u_i(x^T)\}$$

(3)

This game indeed has a huge set of Nash-equilibrium payoffs, containing all β individually rational payoffs, or all feasible payoff vectors $(\lambda_1, \ldots, \lambda_n)$ such that $\beta_i \le \lambda_i$, for all $i = 1, \ldots, n$. This result is known as the first "folk theorem." Its proof involves many cumbersome technicalities (one must, for instance, take into account strategy n-tuples where the limits in (3) do not converge). The second and more-important folk theorem (proved in Rubinstein [1979]) analyzes the coalitional equilibria in the repeated game with payoff (3). This game has many strong equilibria, the payoffs of which cover the β core of the original game.

Definition 2.

Given an N-normal-form game $G = (X_i, u_i, i = 1, \ldots, n)$, the β core of G is the subset of those outcomes x^* such that

for all coalition T, $T \subset N$, there exists a joint strategy $x_{T^c} \in X_{T^c}$ such that for all joint strategy $x_T \in X_T$

$$No \begin{cases} u_i(x_T, x_{T^c}) \geq u_i(x^*) & \text{for all } i \in T \\ \\ u_i(x_T, x_{T^c}) > u_i(x^*) & \text{for at least one } i \in T \end{cases}$$

The stability property of an outcome in the β core is stronger than that of the α core. A deviating coalition T can be countered by the complement coalition T^c even if the members of T keep their joint strategy x_T secret. The counter-threat x_{T^c} is <u>blind</u>, i.e., independent of x_T. Of course, the fact that T is plotting a switch from the agreed-upon strategy $x^*_{T^c}$ is public, otherwise the players in T^c are not aware of the betrayal and cannot counter it. Clearly, the β core is a subset of the α core and contains all strong equilibrium outcomes.

$$SE(G) \subset C_\beta(G) \subset C_\alpha(G)$$

The reader interested in a precise statement of the folk theorems should consult the survey by Aumann [1981].

The economic applications of these results (especially to the theory of imperfect competition) are covered in Friedman [1984].

3. THREATS AND TACTICAL EXCHANGE OF INFORMATION

In two-person games, we can compare various deterring scenarios by their respective credibility and/or feasibility. For instance, a passive threat (in a strong equilibrium) is easier to implement than an active one, or a blind threat is more feasible than one which is not (see Definition 2). The issue of credibility is described by the concept of warnings. The concept involves threats that coincide with a player's best reply (see below). As the classification result demonstrates (Theorem 1 below), these distinctions have much to do with the tactical exchanges of information analyzed in Chapter 2.

Let $G = (X_1, X_2, u_1, u_2)$ be a finite two-person game (most results below extend to the case of compact strategy sets and continuous payoff functions; they do not extend to games with three players). i) The α core of G is the set of its imputations:

$$x \in C_\alpha \iff \begin{cases} x \text{ is Pareto-optimal} \\ \\ x \text{ is individually rational: } \alpha_i \leq u_i(x) \; i = 1, 2 \end{cases}$$

It is never empty. Its outcomes can be stabilized by threats violating privacy (as in the wolf-sheep lemma; see Chapter 2, Lemma 3). ii) The $\underline{\beta\ \text{core}}$ is the set of its Pareto-optimal outcomes where each player gets at least utility $\beta_i = \inf_{x_j} \sup_{x_i} u_i$:

$$x \in C_\beta \iff \begin{cases} x \text{ is Pareto-optimal} \\ \\ \beta_i \le u_i(x) \quad i = 1, 2 \end{cases}$$

This follows from Definition 2, as can be checked. The β core is thus empty iff <u>the competition for the second move arises</u> (Definition 3, Chapter 2). Its outcomes can be stabilized by blind threats. iii) The $\underline{\text{s core}}$ is, by definition, the set of those imputations x^* that can be stabilized by a deterring scenario (x^*, ξ_1, ξ_2) where ξ_i, $i = 1, 2$ is a <u>warning</u>, or a selection of Player i's best reply correspondence:

$$\forall\ x_j \ne x_j^* \quad \begin{cases} u_j(x_j, \xi_i(x_j)) \le u_j(x^*) \\ \\ u_i(x_j, \xi_i(x_j)) = \sup_{x_i \in X_i} u_i(x_j, x_i) \end{cases}$$

The s core is denoted C_s.

Warnings are credible threats insofar as my reaction to your deviation coincides with what my own preferences suggest. I would use that strategy even if I was not threatening you.

The s core is closely related to the Stackelberg utility levels S_i, $i = 1, 2$ (see (3) (4) Chapter 2) as the following result shows.

Lemma 1.

Suppose u_i is one to one on $X_1 \times X_2$ for $i = 1, 2$. Then, the s core is made up of those Pareto-optimal outcomes where each player gets at least his or her Stackelberg utility level as a leader.

$$
x \in C_s \iff \begin{cases} x \text{ is Pareto-optimal} \\ \\ S_i \leq u_i(x) \quad i = 1, 2 \end{cases}
$$

Therefore the s core is empty iff the competition for the first move arises (Definition 4, Chapter 2).

Proof.

Suppose x* is a Pareto-optimum satisfying $S_i \leq u_i(x^*)$, for i = 1, 2, and denote for all $x_i \neq x_i^*$ the best reply strategy x_j of Player j by $x_j = \xi_j(x_i)$. By definition of S_i we have then

$$
u_i(x_i, \xi_j(x_i)) \leq S_i \leq u_i(x^*)
$$

for all i = 1, 2, and all $x_i \neq x_i^*$. Therefore, (x^*, ξ_1, ξ_2) is a deterring scenario; hence x* is individually rational (by Lemma 1, Chapter 9). We could also invoke the always-true inequality $\sup_{x_i} \inf_{x_j} u_i \leq S_i$. Finally x* is Pareto-optimal by assumption.

Conversely, let x* be an outcome in the s core of G. There exists a deterring scenario (x^*, ξ_1, ξ_2) where $\xi_i(x_j)$

equals the (unique, by the one to one assumption) best reply strategy of Player i to x_j. We prove that x^* satisfies $S_i \leq u_i(x^*)$, $i = 1, 2$. First pick $x_i \neq x_i^*$. Then by the deterring property of ξ_j we have $u_i(x_i, \xi_j(x_i)) \leq u_i(x^*)$. It remains to prove

$$x_j = \xi_j(x_i^*) \Rightarrow u_i(x_i^*, x_j) \leq u_i(x^*)$$

Suppose, on the contrary, $u_i(x^*) < u_i(x_i^*, x_j)$. Because x^* is Pareto-optimal, we get

$$\sup_{y_j} u_j(x_i^*, y_j) = u_j(x_i^*, x_j) < u_j(x^*)$$

a contradiction. <u>QED.</u>

<u>Example 4</u>. <u>A Cournot's quantity-setting duopoly</u> (continued).

We consider a variant of Example 2, Chapter 6 where the cost shows (slightly) increasing or decreasing returns to scale:

$$c_i(x) = \frac{1}{2}x + \varepsilon x^2 \text{ where } \varepsilon \text{ is fixed and small}$$

Hence the normal form:

$$X_1 = X_2 = [0, \tfrac{1}{2}], \quad u_i(x_1, x_2) = x_i(1 - x_1 - x_2) - (\tfrac{1}{2}x_i + \varepsilon x_i^2)$$

We will compute the Stackelberg utility level and check whether the s core is empty or not.

Here u_i is concave in x_i, so the best reply of Player 2 is

$$x_2 = BR_2(x_1) \iff x_2 = \frac{1}{1 + \varepsilon} ((1/4) - (x_1/2)) \simeq (1 - \varepsilon)((1/4) - (x_1/2))$$

As a leader, Player 1 solves

$$S_1 = \max_{x_1} \frac{x_1}{2} - (1 + \varepsilon)x_1^2 - x_1 x_2(x_1) \simeq \frac{1 + \varepsilon}{4} x_1 - (\frac{1 + 3\varepsilon}{2}) x_1^2$$

$$\Rightarrow x_1^* = \frac{1}{4} \frac{1 + \varepsilon}{1 + 3\varepsilon} \Rightarrow S_1 = \frac{1}{32} \frac{(1 + \varepsilon)^2}{1 + 3\varepsilon} \simeq \frac{1}{32} (1 - \varepsilon)$$

To check whether (S_1, S_2) $(S_1 = S_2)$ is a feasible utility vector, we will compute $\max_{x_1, x_2} (u_1 + u_2) = P(\varepsilon)$. Since the game is symmetrical, its s core is <u>nonempty iff</u> $S_1 + S_2 \leq P(\varepsilon)$. In fact, $u_1 + u_2 = \frac{x}{2} - x^2 - \varepsilon(x_1^2 + x_2^2)$, where $x = x_1 + x_2$. Now we distinguish two cases. i) For $\varepsilon \geq 0$ (constant or decreasing returns to scale) $u_1 + u_2$ is a concave function of (x_1, x_2) with a maximum at $x_1 = x_2 = \frac{1}{8}(1 - \varepsilon/2)$ and

$$P(\varepsilon) = \frac{1}{16} (1 - \frac{\varepsilon}{2}) > S_1 + S_2 = \frac{1}{16} (1 - \varepsilon)$$

Thus the s core is nonempty and small. For $\varepsilon = 0$ the pair $(S_1, S_2) = ((1/16), (1/16))$ is a Pareto-optimal utility vector so the s core contains the single outcome $((1/8), (1/8))$ (where joint profit is maximal and equally shared). ii) For $\varepsilon < 0$ (increasing returns to scale) $u_1 + u_2$ is <u>not</u> a concave function, and its maximum over R_+^2 is achieved for $x_1 = 0$ or $x_2 = 0$. Taking $x_2 = 0$, we have

$$P(\varepsilon) = \max_{x_1 \geq 0} (\frac{x_1}{2} - x_1^2 - \varepsilon x_1^2)$$

so the optimal x_1 is $\frac{1}{4(1 + \epsilon)}$ and $P(\epsilon) = \frac{1}{16(1 + \epsilon)} \approx \frac{1}{16} (1 - \epsilon)$.

Thus the first-order approximation for S_i is not enough!

We have to compute <u>exactly</u> the Stackelberg utility levels:

$$S_i(\epsilon) = \max_{x_1} \frac{1 + 2\epsilon}{4(1 + \epsilon)} x_1 - (\frac{1 + 4\epsilon + 2\epsilon^2}{2(1 + \epsilon)}) x_1^2$$

$$\Longrightarrow S_i(\epsilon) = \frac{1}{32} \frac{(1 + 2\epsilon)^2}{(1 + \epsilon)(1 + 4\epsilon + 2\epsilon^2)}$$

As $(1 + 2\epsilon)^2/(1 + 4\epsilon + 2\epsilon^2) > 1$, this gives $S_1(\epsilon) + S_2(\epsilon) > 1/16(1 + \epsilon) = P(\epsilon)$. So, the s core <u>is</u> empty.

From Lemma 2, Chapter 2, it follows that the β core C_β and the s core C_s cannot both be empty. The competition for the first and second move cannot arise together. We can say more.

<u>*Theorem 1*</u> *(Moulin [1981]).*

The s core and the β core cannot both be empty. If they are both nonempty, their intersection is nonempty too.

<u>*Corollary.*</u>

Two-person games are partitioned into three classes.

i) $C_s \neq \emptyset$ and $C_\beta = \emptyset$: competition for the second move.

ii) $C_\beta \neq \emptyset$ and $C_s = \emptyset$: competition for the first move.

iii) $C_\beta \cap C_s \neq \emptyset$: any imputation in $C_\beta \cap C_s$ can be stabilized by warnings and/or blind threats.

<u>Proof.</u>

Only the second statement needs a proof. Suppose C_s and C_β

are both nonempty and consider the relative position of (S_1, S_2) and (β_1, β_2). If $(S_1, S_2) \leq (\beta_1, \beta_2)$ or $(\beta_1, \beta_2) \leq (S_1, S_2)$, one of C_s, C_β is contained in the other and the claim is obvious.

It remains to consider the case where, say, $\beta_1 < S_1$ and $S_2 < \beta_2$. By the finiteness of our game, a 1-Stackelberg equilibrium (Chapter 1) exists; say it is x*. It is such that $u_1(x^*) = S_1$ and $x^*_2 \in BR_2(x^*_1) \Rightarrow u_2(x^*_1, x^*_2) \geq \beta_2$. Hence, the set $A = \{x \in X_1 \times X_2 / S_1 \leq u_1(x), \beta_2 \leq u_2(x)\}$ is nonempty. Any Pareto-optimal outcome in A is Pareto-optimal in $X_1 \times X_2$ as well and is an imputation (since $\alpha_1 \leq S_1$ and $\alpha_2 \leq \beta_2$). Such an outcome belongs to $C_s \cap C_\beta$. <u>QED</u>.

A typical element in class iii) is the prisoner's dilemma, where $C_s \cap C_\beta$ coincides with C_α, the set of imputations. Another example is the duopoly model of Example 4 with $\varepsilon = 0$, where C_s is a singleton whereas $C_\beta = C_\alpha$ contains all individually rational outcomes. Exercises 6, 7, give more details about the classification.

4. EXERCISES

a) <u>Strong equilibrium and α core</u>

1) Two shopowners choose the location of their respective shops along the [0, 1] interval. They supply complementary goods (say sports equipment and travel-agent services), implying a positive externality from either shop to the

other. Moreover Player 1 inclines to be located as close to zero as possible, whereas Player 2 seeks to be located as far as possible from zero.

Specifically, they face the following game.

$$X_1 = X_2 = [0, 1]$$
$$u_1(x_1, x_2) = \alpha_1 x_1 - |x_1 - x_2|$$
$$u_2(x_1, x_2) = \alpha_2(x_2 - 1) - |x_1 - x_2|$$
$$\text{where } \alpha_1 < 0 < \alpha_2$$

Suppose that $|\alpha_i| \leq 1$, $i = 1, 2$. a) Compute the best-reply correspondences of both players and the Nash equilibrium. Discuss their stability. b) Find the strong equilibrium outcomes.

2) A three-player prisoners' dilemma

Each player can be aggressive (A) or cooperative (C) and the game is symmetrical. The various payoffs are listed below.

(C, C, C) payoff vector (2, 2, 2)

(A, C, C) payoff vector (3, 1, 1)

(A, A, C) payoff vector (2, 2, 0)

(A, A, A) payoff vector (1, 1, 1)

For instance, the three players are three firms competing by setting a regular price (C) or using a dumping policy (A). The maximal joint profit is 6 (when all players are cooperative)

and decreases by one unit per aggressive player. Each player
switching from C to A gains one additional unit of profit
and decreases by one the profit of the other two.

Prove that the dominating strategy equilibrium is Pareto
dominated and that the game has no strong equilibrium. Prove
that there are exactly four imputations and that the α core
coincides with the set of imputations.

3) A symmetrical three-player game

Each player must pick one among the three players, possibly
himself. Hence $X_1 = X_2 = X_3 = \{1, 2, 3\}$. The payoffs of
the game are deduced from those listed below by the symmetrical
character of our game.

$$
\begin{aligned}
(x_1, x_2, x_3) &= (1, 2, 3) \text{ payoffs:} &(0, 0, 0) \\
&= (1, 2, 1) \text{ payoffs:} &(0, 0, -1) \\
&= (1, 3, 1) \text{ payoffs:} &(0, 2, 0) \\
&= (1, 1, 1) \text{ payoffs:} &(3, 1, 1) \\
&= (1, 3, 2) \text{ payoffs:} &(0, 2, 2) \\
&= (2, 3, 1) \text{ payoffs:} &(2, 2, 2) \\
&= (2, 3, 2) \text{ payoffs:} &(-1, 3, 3)
\end{aligned}
$$

a) Prove that none of the Nash equilibria is Pareto-optimal.
Thus the game has no strong equilibrium. b) Describe the
imputations (five outcomes) and the α core (two outcomes).

b) Repeated games

4) Repetition of the prisoner's dilemma

In the prisoner's dilemma (Example 2, Chapter 3) repeated over time, we assume that each player uses a stationary strategy with length 1 memory. Thus, Player i's strategy is a triple $(x_i; y_i, z_i)$ where x_i, y_i, z_i all belong to $\{A, P\}$, to be interpreted as follows. Player i plays $x_i = x_i^1$ in the first occurrence of the game (time t = 1). At time $t \geq 2$, he plays y_i if Player j was peaceful at time t - 1, and he plays z_i if Player j was aggressive.

$$x_i^t = y_i \quad \text{if } y_j^{t-1} = P$$

$$x_i^t = z_i \quad \text{if } x_j^{t-1} = A$$

For instance "tit for tat" is (P; P, A). The overall payoff is the Cesaro mean limit of instantaeous payoffs. a) Retain one copy of each pair of equivalent strategies and write the remaining 6 x 6 bimatrix game. b) Prove this game is dominance solvable and find its sophisticated equilibrium.

c) β core and s core in two-person games

5) A two-person game $X_1 = X_2 = R$

$$u_1(x_1, x_2) = - \{x_1^2 + 2ax_1x_2 + x_2^2 - 2x_1 - 2ax_2\}$$

$$u_2(x_1, x_2) = - \{x_1^2 + 2ax_1x_2 + x_2^2 - 2ax_1 - 2x_2\}$$

where a is a real parameter such that $|a| < 1$. a) Compute
the Pareto-optimal outcomes and their associated utility
vectors. b) Compute each player's optimal utility vector
as a leader. Is the s core empty? If not, compute it. In
Exercises 6 and 7, G is a finite two-person game whose utility
functions are one to one on X_{12}.

6) <u>More on the classification of two-person games</u>
   a) If G is in Class ii) prove

$$\beta_i < S_i \quad i = 1, 2$$

b) If each player has a (strictly) dominating strategy, then G
is in Class iii). c) If (S_1, S_2) is a Pareto-optimal payoff
vector, then G is in Class iii).

7) <u>Guaranteed deterring scenarios</u>
   Let (x^*, ξ_1, ξ_2) be a deterring scenario of G. We say that
it is <u>guaranteed</u> if no player can suffer a loss by carrying out
his threat:

$$u_1(x_1, \xi_2(x_1)) \leq u_1(x^*) \leq u_1(\xi_1(x_2), x_2)$$

for all x_1, x_2

$$u_2(\xi_1(x_2), x_2) \leq u_2(x^*) \leq u_2(x_1, \xi_2(x_1))$$

The g core of G is the subset, denoted C_g, of those outcomes
x^* in the α core such that there exists at least one guaranteed
deterring scenario (x^*, ξ_1, ξ_2). a) Prove the following

equivalence $x \in C_g <=> [x$ is Pareto-optimal, and
$u_i(x) \leq \beta_i]$.   b) Prove that the g core is a — possibly
empty — subset of the s core:

$$C_g \subset C_s$$

c) If G is in Class i), prove by examples that the g core
can be empty or nonempty.   d) If G is in Class iii), prove
that <u>either</u> its g core is empty, <u>or</u> $\beta = (\beta_1, \beta_2)$ is the
payoff vector of at least one Pareto-optimal outcome.   In
the latter case, prove that $C_g = C_s = C_\beta = (u_1, u_2)^{-1}(\beta)$.

REFERENCES

ARON, R. 1962. Paix et guerre entre les nations. Paris: Calman-Levy Ed.

AUMANN, R.J. 1959. "Acceptable Points in General Cooperative n-Person Games," in: Tucker, A.W., and Luce, R.D., eds., Contributions to the Theory of Games, Vol. IV, Princeton: Princeton University Press, 287-324.

AUMANN, R.J. 1974. "Subjectivity and correlation in randomized strategies". Journal of Mathematical Economics 1: 67-96.

AUMANN, R.J. 1976. Lecture on game theory. Stanford: Standford University, IMSSS.

AUMANN, R.J. 1981. "Survey of Repeated Games," in Essays in Game Theory and Mathematical Economics, by R.J. Aumann et al., Zurich: Bibliographisches Institüt, 11-42.

AUMANN, R.J. and M. MASCHLER. 1964. The bargaining set for cooperative games. Advances in Game Theory. Annals of Math. Studies nº 52. Princeton N.J.: Princeton University Press.

BENOIT, J.P. and V. KRISHNA. 1984. Finitely Repeated Games. Columbia University. Mimeo.

BONDAVERA, O.N. 1962. Teoriia idra v igre n lits (The theory of the Core in an n-person game). Vestnik Leningradskogo Universiteta Seriia Matematika, Mekhanika i Astronomii nº 13: 141-142.

BRAMS, S. 1975. Game theory and politics. London: Glencoe Free Press Collier, Macmillan Pub.

BRAMS, S. and S. MERRILL, III. 1983. Equilibrium strategies for Final-Offer Arbitration: there is no redian convergence. Management Science, 29, 927-941.

CASE, J.H. 1979. Economics and the competitive process. New York: New York University Press.

CAZELLE, P. 1969. "Y a-t-il une science des décisions?" La Nouvelle Critique n° 24: 21-27 et n° 25: 69-75.

CROZIER, M. and F. FRIEDBERG. 1977. L'acteur et le système. Paris: ed. Seuil.

DURKHEIM, E. 1893. De la division du travail social. Paris.

FARQHARSON, R. 1969. Theory of voting. New Haven: Yale University Press.

FREUND, J. 1967. Les théories des sciences humaines. Paris: Presses Universitaires de France.

FRIEDMAN, J.H. 1984. Game theory with applications to economics, mimeo, Virginia Polytechnic Institute and State University.

GABAY, D. and H. MOULIN. 1980. "On the uniqueness and stability of Nash equilibrium in non cooperative games" in Applied stochastic control in econometric and management science, Bensoussan, Kleindorfer, Tapiero Eds., Amsterdam: North-Holland Publishing Co.

GALE, D. 1974. A curious Nim-type game. American Mathematical Monthly, 87, 876-879.

GALE, D. and A. NEYMAN. 1980. Nim-type games. Mimeo. University of California, Berkeley.

GERARD-VARET, L.A. and H. MOULIN. 1978. "Correlation and duopoly", Journal of Economic Theory, 1: 123-149.

GLICKSBERG, I.L. 1952. "A Further Generalization of the Kakutani Fixed Point Theorem, with Application to Nash Equilibrium Points," Proceedings of the American Mathematical Society, 3, 170-174.

GRANGER, G. 1955. Méthodologie économique. Paris: Presses Universitaires de France.

GRETLEIN, R. 1983. "Dominance elimination procedures on finite alternative games". International Journal of Game Theory, 12, 2, 107-114.

GROTE, J. 1974. "A global theory of games". Journal of Mathematical Economics 1, 3: 223-236.

ICHIISHI, T. 1983. "Non-Cooperation and Cooperation," in: Deistler, M., and Schwödiauer, G., eds., Oskar Morgenstern Symposium Proceedings, Vienna/Wurzburg: Physica-Verlag.

KAKUTANI, S. 1941. A generalization of Brouwer's fixed point theorem, Duke Mathematical Journal 8: 457-458.

KUHN, H.W. 1953. "Extensive Games and Problems of Information," in: Contributions to the Theory of Games II, edited by H.W. Kuhn and A.W. Tucker, Princeton: Princeton University Press, 193-216.

LAFFOND, G. and H. MOULIN. 1981. Stability by threats and counter-threats in normal form games, in: Mathematical Techniques of Optimization, Control and Decision, Aubin, Bensoussan, Ekeland, eds., Boston: Birkhaüser.

LEMKE, C.E. and J.T. HOWSON. 1964. "Equilibrium Points of Bi-Matrix Games," SIAM Journal of Applied Mathematics, 12, 413-423.

LUCE, R.D. and H. RAIFFA. 1957. Games and decisions. New York: J. Wiley and Sons.

MASKIN, E. 1983. The theory of Implementation in Nash equilibrium: a survey, forthcoming in Social Goals and Social Organization: a volume in memory of E. Pazner (Schmeidler and Sounenschein eds.).

MAYNARD SMITH, J. 1974. The theory of Games and the Evolution of Animal Conflicts. Journal of Theoretical Biology, 47, 209-221.

MERTENS, J.F. 1983. The Minmax Theorem for U.S.C.-L.S.C. payoff functions, mimeo, Stanford University.

MOULIN, H. 1976. Cooperation in Mixed Equilibrium. Mathematics of Operations Research, 1, 3, 373-386.

MOULIN, H. 1979. Dominance-solvable voting schemes. Econometrica, 47, 6, 1337-1351.

MOULIN, H. 1979. "Two and three person games: a local study" in International Journal of Game Theory, 8, 2: 81-107.

MOULIN, H. 1981. "Deterrence and cooperation". European Economic Review, 15: 179-193.

MOULIN, H. 1983. The Strategy of Social Choice, Advanced Textbooks in Economics, North-Holland, Amsterdam.

MOULIN, H. 1984. Dominance-Solvability and Cournot Stability. Mathematical Social Sciences, 7, 83-102.

MOULIN, H. and J.P. VIAL. 1978. "Strategically zero-sum games". International Journal of Game Theory, 7, 3/4: 201-221.

MUSASHI. 1981. The book of five rings. Holden Day: San Francisco.

NASH, J.F., JR. 1951. Non-cooperative games, Ann. Math. 54: 286-295.

NISHIMURA, K. and J. FRIEDMAN. 1981. Existence of Nash Equilibrium in n person games without quasi-concavity, International Economic Review, 22, 3, 637-648.

NOVSHEK, W. 1983. On the existence of Cournot Equilibrium, forthcoming Review of Economic Studies.

O'NEILL, B. 1985. International Escalation and the Dollar Auction, mimeo, Northwestern University.

ORTEGA, J.M. and W.C. RHEINBOLDT. 1970. Iterative solution of non-linear equations in several variables. New York: Academic Press.

OWEN, G. 1982. Game Theory (2nd edition), new York: Acacemic Press.

PARTHASARATHY, T. and T.E.S. RAGHAVAN. 1971. Some topics in two-person games. New York: American Elsevier.

PELEG, B. 1984. Game theoretic analysis of voting in committees, Boston, Cambridge University Press.

RAND, D. 1978. "Exotic phenomena in games and duopoly models". Journal of Mathematical Economics 5, 2: 173-184.

RIVES, N.W. 1975. "On the history of the mathematical theory of games". Hope 7, 4.

ROCHET, J.C. 1980. "Selection of a unique equilibrium payoff for extensive games with perfect information", mimeo, Université Paris IX.

ROSEN, J.B. 1965. "Existence and uniqueness of equilibrium points for concave N-person games". Econometrica 33: 520-533.

ROSENTHAL, R. 1972. "Cooperative games in effectiveness form". Journal of Economic Theory 5, 1.

ROUSSEAU, J.J. 1755. Discours sur l'origine des inégalités, Paris.

RUBINSTEIN, A. 1979. "Equilibrium in supergames with the over-taking criterion". Journal of Economic Theory 21, 1: 1-9.

RUBINSTEIN A. 1982. "Perfect Equilibrium in a Bargaining Model," Econometrica 50, 97-109.

RUCKLE, W.H. 1982. A Division Game, or How far can you trust mathematical induction, mimeo, Clemson University.

SCARF, H. 1967. The core of an N-person game. Econometrica 35: 50-69.

SCHELLING, T.C. 1971. The strategy of conflict. Cambridge: Harvard University Press, 2nd Ed.

SCHELLING, T.C. 1979. Micromotives and macro-behaviour. New York: Norton Publ.

SCHMEIDLER, D. 1969. The Nucleolus of a Characteristic Function Game. SIAM Journal on Applied Maths. 17, 1163-1170.

SCHOTTER, A. and G. SCHWÖDIAUER. 1980. "Economics and game theory: a survey." Journal of Economic Litterature 18, 2.

SELTEN. R. 1975. Reexamination of the perfectness concept for equilibrium points in extensive games, Int. J. of Game Theory, 4, 25-55.

SHAPLEY, L.S. 1971. "Cores of Convex Games," International Journal of Game Theory, 1, 11-26.

SHERER, F.M. 1970. Industrial pricing. Chicago: Rand MacNally.

SION, M. and P. WOLFE. 1957. "On a game without a value". Contributions to the theory of games, vol. 3. Annals of Maths. Studies 39. Princeton: Princeton University Press.

TIJS, S.H. 1981. "Nash equilibria for non-cooperative n-person games in normal form". SIAM Journal Appl. Math. 23, 2: 225-237.

YOUNG, H.P., N. OKADA, and T. HASHIMOTO. 1982. Cost allocation in water resources development, Water Resources Research 18, 463-475.

AUTHOR INDEX

Aron: 211
Aumann: 192, 195, 230, 242, 253

Benoit: 249
Bertrand: 123
Bondavera: 228
Borda: 53
Borel: 19, 29, 70
Brams: 4, 29

Case: 65, 66, 145
Cazelle: 5
Choquet: 34
Crozier: 4

Durkheim: 5

Farqharson: 4, 73
Freund: 10
Friedberg: 4
Friedman: 116, 254

Gabay: 142
Gale: 21, 23, 25, 26
Gerard-Varet: 207
Glicksberg: 178
Granger: 5
Gretlein: 101
Grote: 117

Hayek: 5

Ichi-Ishi: 246

Kakutani: 115
Krishna: 249
Kuhn: 21, 24
Kuhn's reduction algorithm: 84

275

SUBJECT INDEX